IV opAmp FB Resistor

$$R_F = \frac{V_O}{I_i}$$

V_O ← MAX Voltage out of opamp you want

I_i ← MAX current coming from DAC

Electronic Circuit Guidebook

Volume 3:

Op Amps

Electronic Circuit Guidebook

Volume 3:
Op Amps

Written By
Joseph J. Carr

PROMPT®
PUBLICATIONS

A Division of Howard W. Sams & Company
A Bell Atlantic Company

PROMPT© Publications is an imprint of Howard W. Sams & Company, A Bell Atlantic Company, 2647 Waterfront Parkway, E. Dr., Indianapolis, IN 46214-2041.

International Standard Book Number: 0-7906-1131-7

Library of Congress Catalog Card Number: 97-65787

Acquisitions Editor: Candace M. Hall
Editor: Natalie F. Harris
Assistant Editors: Pat Brady, Loretta Leisure
Production Support: Shey Query, Karen Mittelstadt
Typesetting: Natalie Harris, Loretta Leisure
Indexing: Natalie Harris
Cover Design: Suzanne Lincoln, Christy Pierce
Graphics Conversion: Terry Varvel
Illustrations and Other Materials: Courtesy of the Author

PRINTED IN THE UNITED STATES OF AMERICA

9 8 7 6 5 4 3 2 1

Contents

Preface 5

Chapter 1
Linear IC Amplifiers 3

Amplifier Fundamentals 5
Operational Amplifiers 8
Single Supply Operation 11
Grades and Classes of Op Amp 14
Current Difference ("Norton") Amplifiers 17
Operational Transconductance Amplifiers (OTA) 18
Protecting Linear IC Amplifiers 18
Common IC Package Types 21
Conclusion 24

Chapter 2
The Ideal Operational Amplifier 25

Properties of the Ideal Op Amp 26
Conclusion 32

Chapter 3

Inverting Follower Op Amp Circuits 35

What's a "Virtual" Ground? 35
Developing the "Transfer Equation" For the Inverting Follower 36
Frequency Response Tailoring 39
AC-Coupled Inverting Followers 41
Multiple-Input Inverting Followers 42
Basic Design Procedure 43
Project Examples 45

Chapter 4

Noninverting Followers 49

Application of Unity Gain Followers 50
Noninverting Followers With Gain 53
Advantages of Noninverting Followers 54
AC-Coupled Noninverting Amplifiers 55
Designer's Casebook Example 60
Transfer-Couple Noninverting Amplifier 63
Conclusion 64

Chapter 5

Ideal vs. Practical Operational Amplifiers 67

DC Offset Problems 68
Corrections for DC Offset Problems 70
Frequency Compensation Problems 75

Chapter 6

DC Differential Operational Amplifier Circuits 83

Common Mode Signals 83

Differential Signals 85
Differential Input Stage 86
Practical Circuits 87
Designer's Casebook Example 88
CMRR Adjustment 90
Universal Diff-Amp Project 92
Variable Gain Differential Amplifier 93
Some Special Diff-Amps 94
In the Next Chapter 95

Chapter 7
Instrumentation Amplifiers 83

Simple IA Circuit 84
Standard Instrumentation Amplifiers 85
Gain Equation For the IA Circuit 86
Gain Control For the IA 86
Common Mode Rejection (CMRR) Adjustment 88
AC Instrumentation Amplifiers 89
An ECG Amplifier 92
Some Commercial IC IA Examples 96
Next Chapter 99

Chapter 8
Isolation Amplifiers 101

Battery-Powered Designs 103
Carrier-Operated 103
Optically-Coupled Circuits 106
Current Loading Methods 108
Some Commercial Examples 109
Conclusion 113

Chapter 9

Operational Transconductance Amplifiers 129

Voltage Amplifier From the OTA 133
Next Chapter 134

Chapter 10

Current Difference Amplifiers 135

CDA Configuration 136
Inverting Follower Circuits 138
Designer's Casebook Example 141
Noninverting Amplifier Circuits 143
Super-Gain Amplifier 145
Differential Amplifiers 146
AC Mixer/Summer Circuits 148
Differential Output 600-Ohm Line Driver Amplifier 148
Using Bipolar DC Power Supplies 150

Chapter 11

Audio Integrated Circuits 151

Audio Preamplifiers 151
DC Power Connections 152
Special Purpose IC Amplifiers 152
Operational Amplifier Preamps 154
Tone Control Circuits 159
Audio Mixer Circuits 162
600 Ohm Audio Circuits 165
Some Practical Preamplifier Circuits 167
Audio Compression Amplifier 169
IC Audio Power Amplifiers 170
The LM-386 Audio Preamplifier/Power Amplifier 170

Chapter 12
Wideband and Video Linear IC Amplifiers 175
General Wideband Preamplifier 175
MAR-1 Wideband Preamplifier 177
Wideband Op Amps and OTAs 180
Voltage Controlled Wideband Amplifier 180
Wideband Amplifier Construction Hints 182

Chapter 13
Active Analog Filters 185
Filter Frequency Response Characteristics 185
High-Pass and Low-Pass Filter Designs 190
For Further Reading 195

Chapter 14
Waveform Generators and Timer Circuits 197
Relaxation Oscillators 198
Negative Resistance Devices 200
Feedback Oscillators 201
Sine Wave Oscillators 203
Wein Bridge Oscillators 203
Twin-Tee Oscillators 205
Bridged-Tee Oscillators 205
Phase Shift Oscillators 206
Non-Sinusoidal Waveform Generators 207
Square Wave Generators 208
Square Waves from Sine Waves 211
Sawtooth Generator Circuit 212
Digital Waveform Generators 214

Timer Circuits 217
Conclusion 219

Chapter 15
Instrumentation and Other Circuits 221
Keeping the Circuits Stable 221
Integrators and Differentiators 224
Electronic Integrator Circuits 227
A Practical Integrator 229
Electronic Differentiators 231
Precise Rectifiers 233
Peak Detector Circuits 235
Quadrature Detector 237
An Instrumentation Example 237
Mean Value Detector 240
Minimum Value Detector 241

Chapter 16
DC Power Supplies for Op Amp Circuits 243
DC Power Supplies 243
DC Power Distribution 246
Power Distribution Schemes for Circuit Boards 252
Isolating Circuits on the PCB 253

Index 255

Preface

Electronics is a fascinating field, and has been popular ever since the very early days of radio (early 1900s). The typical electronic and radio hobbyist has always been something of an activist elite who experiments with practical circuits, designs personal projects, and builds everything from simple, "one-evening" jobs to complex systems that take months to complete. Personally, I've built projects from simple crystal set radios to large and complex instruments. I've even seen a large computer-driven X-Y plotter project advertised in the magazines just recently, indicating that hobby construction is still very popular. This book is designed for electronic hobbyists and others who enjoy learning about or building fundamental electronic circuits.

At one time all electronic projects were composed of discrete components such as resistors, capacitors, transistors and vacuum tubes. In the early 1960s, a great change — what some today would call a paradigm shift — in electronic construction occurred with the advent of integrated circuitry. The first commercial ICs (integrated circuits), devices such as the μA-709 operational amplifier and the μA-703 FM IF amplifier that elicited excitement in their day, are now considered not only over the hill, but out of sight behind us. These devices are small-scale integration and extremely primitive compared with products sold today. However, in 1967, the early devices were the latest and greatest — with prices commensurate with that status.

Since those early days, we have progressed through several orders of integration, where we now commonly see medium-scale integration (MSI), large-scale integration (LSI) and very large-scale integration (VLSI) devices sold through nonindustrial distributors and hobbyist-oriented parts stores such as Radio Shack. Now on the market are monolithic microwave integrated circuit (MMIC) devices, such as the Mini-Circuits Laboratories MAR-xx series.

The purpose of this book is to give you some insight into how practical linear ICs work in actual, practical real-life circuits. Because of their widespread popularity, operational amplifiers (op amps) figure heavily in the presentation, although other types of amplifiers are not overlooked. The idea is to allow you to design and configure your own circuits, and the material is presented with that goal in mind.

Whatever else this book is, it is intended to be a practical workshop aid. Although most authors probably like to see their works given a place of honor in their reader's library, on the bookshelf in the living room for instance, it is my desire that this book be kept on your workbench, get folded over from long and hard use, and (of course) amply solder splashed.

Chapter 1

Linear IC Amplifiers

In 1966 I worked as an electronics technician repairing stereo amplifiers and receivers. One day, while working on an H.H. Scott receiver (one of the better brands in those days), I came across a series of semiconductor components in the FM IF stage that had a funny look about them. They looked like plastic epoxy transistors with six "dead bug" legs sticking up in the air. At first perplexed, my mind snapped into gear as it remembered an article on the then-new "integrated circuits" (ICs). What I was looking at was one of the earliest examples of a µA-703 IF/RF gain block IC device. Shortly thereafter, I discovered the first IC operational amplifier, the µA-709 (which cost $80 then, but goes five for a buck now when you can find them. The µA-709 was truly revolutionary in its day).

There are few who would argue that the integrated circuit has revolutionized the entire electronics industry. Virtually unknown a couple decades ago, the IC "chip" brought us both the extreme miniaturization of electronics circuits, and previously unheard-of component densities. It is unlikely that the miracles of modern electronics — from "smart" cardiac pacemakers to the moon landing program, from consumer entertainment electronics to the most massive fifth-generation super computers — would have been possible without the little chip.

Device performance was also improved by IC construction. Consider thermal drift of DC amplifiers, for example. Even low-cost modern IC operational amplifiers are several orders of magnitude better off than traditional vacuum tube or discreet transistor models because all of the semiconductors and internal resistors (the main sources of drift) share the same thermal environment in ICs. In discrete circuits, those components are spread out over several square inches of circuit board and thus do not share the same thermal environment — so drift is more pronounced than in the IC version of the circuit.

Cost is another great advantage of the integrated circuit. Early transistor and vacuum tube operational amplifiers were not only larger and ran hotter than their modern IC counterparts, but they were more costly as well. In addition, those early amplifiers only poorly approximated the performance of the ideal textbook version of the amplifier (see Chapter 2). Modern IC operational amplifiers come so close to the ideal (especially premium types) that textbook authors no longer look like inveterate liars to those students who try to translate the theory into workbench practice (sigh).

The first ICs appeared in the mid-1960s, and were low-density devices (for example, the μA-703 mentioned earlier). As industry learned their trade, the number of active devices on the chip skyrocketed. Today, medium scale integration (MSI) and large scale integration (LSI) are the norm — and super-scale integration devices loom on the horizon as the frontiers of technology crumble to the onslaught of the quantum cooks of Silicon Valley.

The principal subject of the first section of this book is the operational amplifier. While our overall topic is ICs in general, op amps loom large and are the most useful to most readers. We will also consider digital devices and other linear analog devices, such as the operational transconductance amplifier (OTA) and current difference amplifier (CDA, also called "Norton Amplifier"). First, let's take a short look at amplifier basics.

Amplifier Fundamentals

Amplifiers are classified according to several schemes: conduction angle, transfer function and so forth. The conduction angle method of classification places the amplifier into one of several different classes labeled A, B, C and AB (which is a combination class of A and B type characteristics, and is further subdivided into AB1 and AB2). The different classes are divided according to the conduction angle of the active device, be it tube, transistor or whatever. The conduction angle is the number of degrees of a sine wave input signal over which output current flows. In a transistor amplifier, for example, we examine the collector current to determine classification. The requirements for the various classes are given below:

Class-A. Output current flows over all 360 degrees of the input sine wave cycle. This class of amplifiers is said to be "linear" because the output waveform is merely a larger (i.e., amplified) version of its input waveform. Typical Class-A amplifiers conduct current regardless of the input signal level. When the input signal is zero, then the power developed in the output circuit (e.g., transistor collector) is dissipated as heat rather than as signal applied to the load. Class-A amplifiers are often used as power amplifiers in low-cost consumer audio and radio receiver circuits. Mostly, however, Class-A amplifiers are used as low-level current and voltage amplifiers or audio preamplifiers.

Class-B. In this class of amplifier output current is conducted over 180 degrees (i.e., one-half) of the input sine wave cycle. Thus, only fifty percent of the input signal is reproduced in the output load circuit. A single-ended Class-B amplifier is therefore nonlinear (i.e., the output signal is distorted), but can be made linear if two Class-B amplifiers drive the same load 180 degrees out of phase. Typically, the two amplifier outputs will be connected to the load in parallel, and their respective input signals are derived from 180 degree out-of-phase sources. Such an amplifier is said to be a "Class-B push-pull amplifier." Although more hardware is needed by Class-B amplifiers when operating in the linear mode, compared with Class-A amplifiers, they are also a lot more efficient with respect to the consumption of DC power. When used as a linear power amplifier, Class-A amplifiers are typically 30 to 40 percent efficient (i.e., only 30 to 40 percent of DC power consumed is delivered as audio to the load, and 60 to 70 percent is converted to heat). The Class-B amplifier, on the other hand, operates at efficiencies of 50 to more than 70 percent.

Class-AB. This combination class provided some of the better attributes of Class-A and Class-B amplifier circuits. Output current for flows more than 180 degrees of the input sine wave cycle, but less than the full 360 degrees. When two stages are combined in a push-pull configuration, the Class-AB amplifier is capable of linear operation at better efficiency than Class-A. In subclass AB1, no base current flows in the transistor on signal peaks, while in subclass AB2 a small amount of base current does flow.

Class-C. In the Class-C amplifier output current flows for considerably less than 180 degrees of the input sine wave cycle (120 degrees is common). This form of amplifier is never linear, but it is capable of very high efficiency. The Class-C amplifier is used only in RF power amplifier circuits in radio transmitters that are either high-level amplitude modulated, frequency modulated (where the distortion of amplitude peaks does not offend the modulation of the transmitter), CW on-off telegraphy or unmodulated (which group includes industrial inductive heating and medical diathermy). Single-sideband and low-level AM modulated radio transmitters cannot use a Class-C amplifier in the signal chain, so will use one of the other classes.

The "transfer function" of any electrical circuit (including amplifiers) is the ratio of the output function over the input signal function. For example, a voltage amplifier transfer function, indicated by the symbol A_v, is the ratio of output voltage over input voltage: V_o/V_{in}. There are several transfer functions used in common amplifier configurations: power, voltage, current, transresistance and transconductance.

The power amplifier transfer function is of the general form:

- $$A = \frac{P_o}{P_{in}} \qquad eq.\ (1\text{-}1)$$

— where:

A is the power gain.
P_o is the output power.
P_{in} is the input power.

Other forms of transfer function relate output power to input current (P_o/I_{in}) or input voltage (P_o/V_{in}). The latter applies to audio amplifiers, for example, in which an input signal voltage (usually expressed in millivolts) drives a loudspeaker load to a given power level (expressed in watts).

The voltage amplifier, which includes most operational amplifiers, has a transfer function that relates output signal voltage to input signal voltage:

$$A_{vol} = \frac{V_o}{V_{in}} \qquad \text{eq. (1-2)}$$

— where:

A_{vol} is the voltage gain.
V_o is the output voltage.
V_{in} is the input voltage.

It is important that input and output functions be expressed in terms of the same units (volts, millivolts, etc.). If the input potential is expressed in millivolts, then we must either convert it to volts or express the output signal in millivolts also. For example, assume that an excitation of 10 mV produces a 2 volt output signal. The gain of this amplifier is correctly expressed as 2/0.010 = 200, not 2/10 = 0.2.

The current amplifier relates the output current to the input current, and has a transfer function of the form:

$$A_i = \frac{I_o}{I_{in}} \qquad \text{eq. (1-3)}$$

— where:

A_i is the current gain.
I_o is the output current.
I_{in} is the input current.

Again, we must remain consistent regarding units (amperes, milliamperes, microamperes, etc.).

The transconductance and transresistance amplifiers offer transfer functions that look a little weird on first glance. While the previous amplifier types had dimensionless transfer functions (volts/volts is dimensionless because the "volts" units cancel out mathematically), the transresistance amplifier has its transfer function expressed in units of resistance. Similarly, the transconductance amplifier has transfer function units in siemens, millisiemens or microsiemens (all units of conductance, the reciprocal of resistance). An older unit for transconductance seen in some op amp books is the mho, which is merely a different name for the siemen (1 mho = 1 siemen). For the transresistance amplifier the transfer function is:

- $$A_r = \frac{V_o}{I_{in}} \qquad \textit{eq. (1-4)}$$

— where:

A_r is the gain in ohms.
V_o is the output potential in volts.
I_{in} is the input current in amperes.

The transconductance amplifier is the inverse of the transresistance amplifier. Recall by Ohm's law, resistance R is the quotient V/I. Conductance (G), is the reciprocal of resistance (G = 1/R), so conductance is expressed by I/V — in the units once called "mhos" but now called "siemens." The transfer function of a transconductance amplifier (of which, we will look at an IC version later on) relates an output current to an input voltage:

- $$A_g = \frac{I_o}{V_o} \qquad \textit{eq. (1-5)}$$

— where:

Ag is the gain in siemens.
Io is the output current in amperes.
Vin is the input signal potential in volts.

Of the various classes of amplifiers, most useful to us in this text are voltage, current and transconductance.

Operational Amplifiers

The operational amplifier (OA) is the most commonly used linear IC amplifier in the world. Originally, the first vacuum tube OAs were designed to calculate

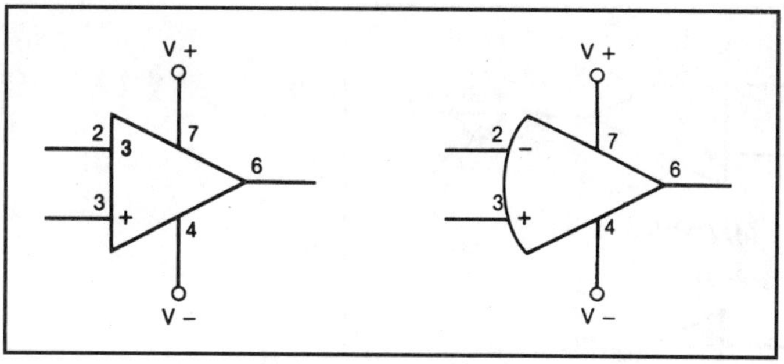

Figure 1-1. Op-amp circuit symbols: A) most commonly used version; B) original version.

mathematical operations in old-fashioned analog computers, hence the name "operational amplifier" sticks today. The IC form of OA today is so widely used that it no longer has any significant role in the all but totally obsolete analog computers of yesteryear. The range of other applications for the OA is, however, truly awesome — it has become a mainstay of more audio, communications, TV, broadcasting, instrumentation, control and measurement circuits than anyone can imagine.

The circuit symbols for the operational amplifier are shown in Figures 1-1A and 1-1B. The symbol shown in Figure 1-1A is by far the most commonly used, and will be used in this book unless otherwise specified. This same symbol could also be used to represent any amplifier with differential inputs, but here will be used as the only proper symbol for op amps.

The symbol shown in Figure 1-1B is also seen occasionally, and some people regard it as the only proper symbol for operational amplifiers. The IEEE standard for circuit symbols, as well as industrial semiconductor companies such as Burr-Brown Corporation, specify the symbol of Figure 1-1B.

There are two input pins on the typical OA. The inverting input (-) produces an output signal that is 180 degrees out of phase with the input signal (see Figure 1-2A). Here we see a positive input signal produces a negative output signal, and vice versa. The noninverting input (+) produces an output signal that is in phase with the input signal (Figure 1-2B). In the various applications shown in this book we will use either or both inputs. An amplifier that uses only the inverting input is called an “inverting follower,” while the amplifier

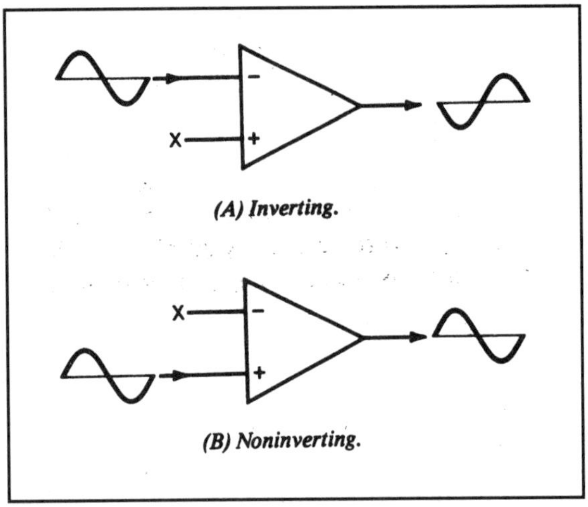

Figure 1-2. Operating modes of an op-amp: A) inverting; B) noninverting.

that uses only the noninverting input is called a "noninverting follower" (why the term "follower" is used, no one under sixty seems to remember — and quite a few over-60 people probably never knew). An amplifier that uses both inputs might be a summer amplifier in some cases, but it is more likely a "differential amplifier." That is, it produces an output signal that is a function of its gain and the difference between two input signal potentials.

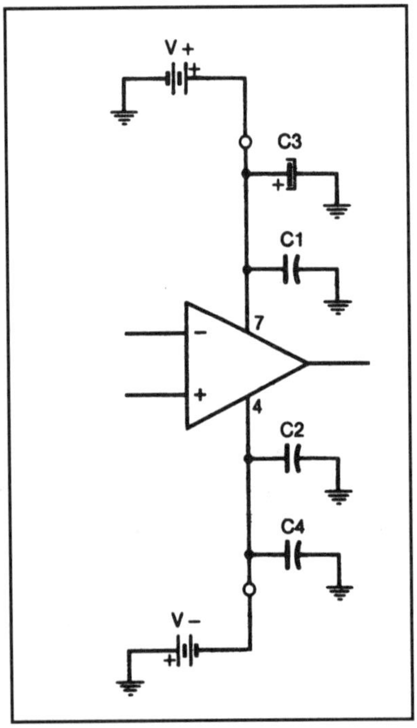

Figure 1-3. Configuration for bipolar DC power supply, with correct decoupling capacitor scheme.

The operational amplifier normally operates from a bipolar DC power supply, such as shown in Figure 1-3 (the pin numbers in this figure are for the so-called "industry standard" 741-device). This circuit (Figure 1-3) shows that the two DC power supplies are independent of each other. The V+ power supply is positive with respect to common (or ground), while the V- supply is negative with respect to the common. The operational amplifier manufacturer will specify minimum and maximum values for V- and V+. Typically, the maximum voltages will be on the order of ±18 volts, with some offering ±22 volts (or in at least one advertised case ±40 volts).

Because most practical circuits operate with equal bipolar DC power supplies, it is also true that most of those using popular operational amplifiers with the 30 volt limit also limit the V- and V+ DC power supplies to not more than 15 volts each.

Single Supply Operation

The operational amplifier is intended for dual or bipolar power supply operation. There are, however, many applications where only one polarity DC power supply is available. In order to operate the op amp in these cases, we must either supply the missing potential, or devise a method for getting around the need for the missing potential.

It is reasonably easy to supply a missing potential. All we need is to add a DC-to-DC converter circuit that provides the needed voltage from the existing voltage. There are quite a few devices on the market that will produce either

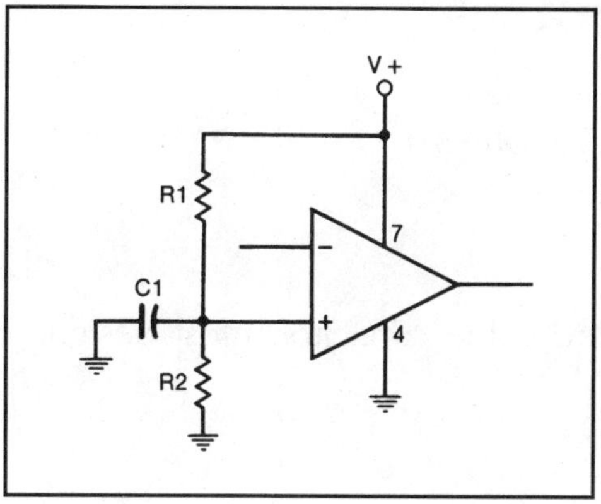

Figure 1-4. Op-amp operation from a single polarity DC supply.

-15 volts from +15 volts (or 12 volts, as the case may be), or will produce isolated ±15 volt potentials from an existing non-isolated +15 volt potential. A popular version of this type of circuit is seen in hobbyist magazines from time to time. It uses a 555 IC timer device operated at or near 100 kHz, and a rectifier-filter circuit to produce a negative voltage. Of course, a voltage regulator can be added if needed. Typically, negative regulators will be in the 79xx or LM-320-xx series.

Another method for using a single DC power supply is shown in Figure 1-4. Here we use a resistor voltage divider, R1/R2, to bias the noninverting input of the operational amplifier to some potential between ground and V+. The V- terminal of the op amp is grounded. The bias voltage on the noninverting input also appears on the output terminal as a DC offset potential. Unless the following circuit somehow doesn't care about this offset potential the output must be capacitor-coupled. The value of the bias voltage is found from:

$$V1 = \frac{(V+)R2}{R1+R2} \qquad \text{eq. (1-6)}$$

The capacitor shown in Figure 1-4 is used to place the noninverting input at or near ground potential for AC signals, while retaining the DC level produced by the resistor voltage divider. This capacitor sometimes causes noisy operation of the op-amp (especially when the circuit ground connection is noisy), so is often omitted in practical circuits. The value of the capacitor is such that it has a capacitive reactance of less than R2/10 at the lowest frequency of operation. For example, if the amplifier is designed to work down to 10 hertz, and the value of R2 is 2200 ohms (a typical value in real circuits), then the value of C1 must be such that it has a reactance of 220 ohms or less at 10 Hz. This requirement evaluates to:

$$C1 = 1{,}000{,}000/(2\pi FX)$$
$$C1 = 1{,}000{,}000/((2)(3.14)(10\text{ Hz})(220\text{ ohms}))$$
$$C1 = 1{,}000{,}000/13{,}816$$
$$C1 = 73\ \mu F$$

Because 100 µF is the next higher standard value capacitor, most designers will select 100 µF for C1 instead of 73 µF.

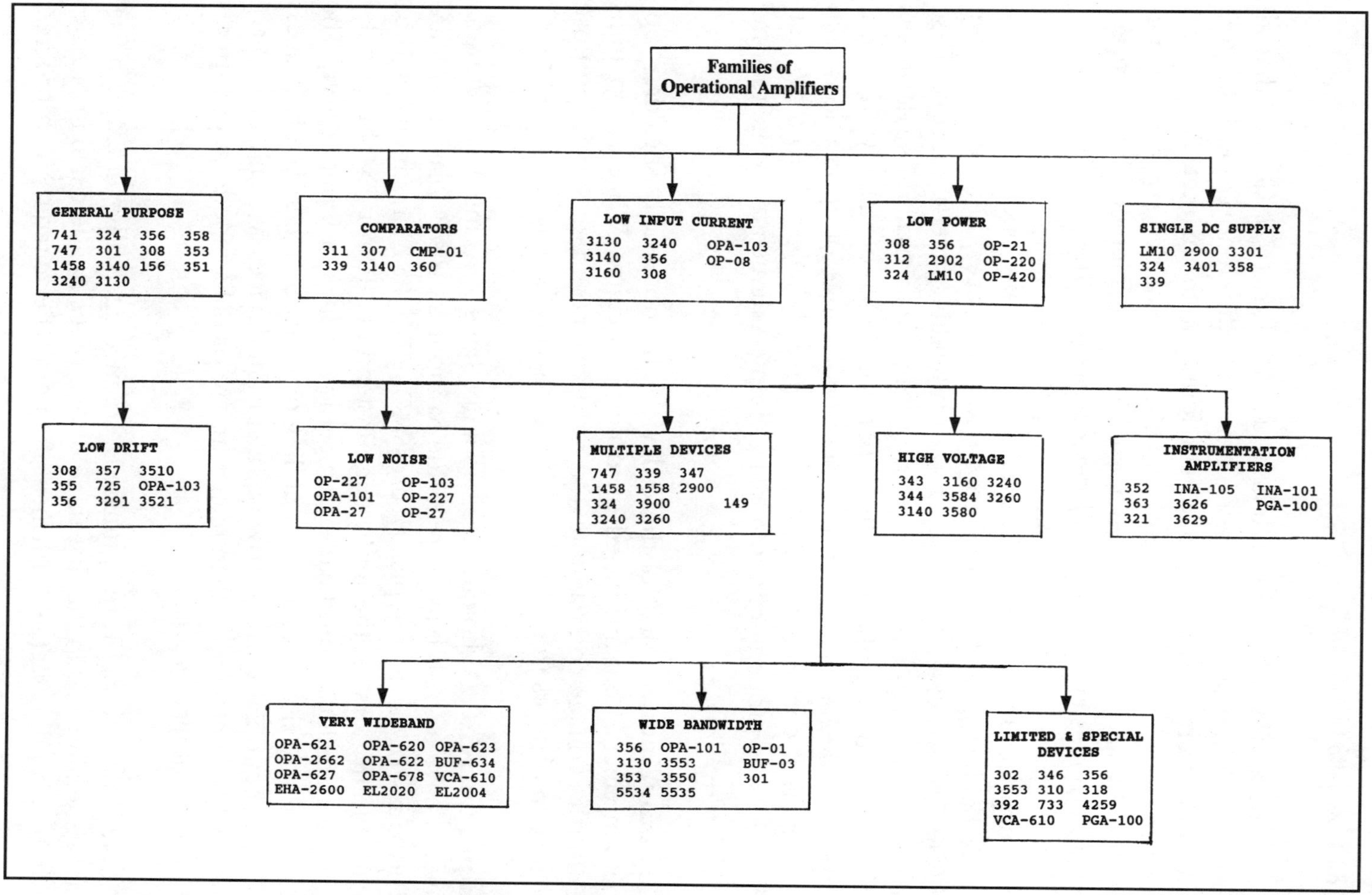

Table 1-1. Grades and classes of op-amps.

Grades and Classes of Op Amp

The term "operational amplifier" refers to a wide variety of devices that all meet the basic definition, but otherwise have very different properties. Table 1-1 shows a hierarchy of devices that can be found on the market. The devices listed are not being recommended against others (the market is always changing), and the listings are not comprehensive. Rather, it is intended to be representative and to illustrate the kinds of devices seen on the market.

General Purpose Op Amps. These are "garden variety" operational amplifiers that are neither special purpose nor premium devices, but they are often limited (the 741, for example, is severely bandwidth limited). Most of these general purpose devices are said to be "frequency compensated," so designers trade-off bandwidth for inherently high, but not absolute, stability. The general purpose devices can be used in a very wide range of applications with very few external components. Op amps are usually selected from this category unless some property of another class brings unique advantage to some particular application.

Voltage Comparators. Most of these devices are not strictly speaking "operational amplifiers," but are based on op amp circuitry. While all op-amps can be used as voltage comparators, the reverse is not true: special purpose IC voltage comparators (e.g., LM-311) cannot usually be used as op amps.

Low Input Current. Although ideal op amps have zero input bias current, real devices have a small bias current due to input transistor biasing or leakage. This class of devices typically uses MOSFET, JFET or superbeta (i.e., Darlington) transistors for the input stage instead of NPN or PNP bipolar devices. Those transducers produce extremely low values of input current. The manufacturer may also elect to use a nulling or selection technique in-process that reduces input bias current. Low input current devices typically have picoampere level currents, rather than microampere or milliampere input currents found in some other devices.

Low Noise. These devices are optimized to reduce internally-generated noise. They are generally used in the first stage or two of a cascade chain in order to improve the overall noise performance of an amplifier circuit. Much progress has been made in the design of low-noise op amps in recent years. Note, however, that some devices, such as the 3140 BiMOS op amp, can be considered almost low-noise when operated at the low end of the DC power supply range (e.g., ±5 VDC), and with a heat sink (if the device is a metal can package).

Low Power. This category of op amp optimizes internal circuitry to reduce power consumption. Many of these devices also operate at very low DC power supply potentials (e.g., ±1.5 VDC).

Low Drift. DC amplifier circuits can experience an erroneous change of output voltage as a function of temperature. Low-drift devices are internally compensated to minimize temperature drift. These devices are typically used in instrumentation circuits where drift is an important concern, especially when handling low level input signals.

Wide Bandwidth. Also called "video" op amps in some literature (especially when the input or output impedance is 50 ohms), these devices have a very high gain-bandwidth product. One device, for example, has a G-B product of 100 MHz, compared with 300 kHz to 1.2 MHz for various 741-family devices. Many of these devices are fixed gain, or unity gain, and only a few (e.g., 5534) are true operational amplifiers.

Very Wide Bandwidth. These amplifiers represent a relatively new entry into the marketplace. These amplifiers are similar to the wide bandwidth, but have true op amp characteristics. Devices in this category have minimum gain-bandwidth products of 30 MHz (so overlap with the above class), but examples are available with gain-bandwidth products of 500 MHz or 650 MHz.

Single DC Supply. These devices are able to provide op amplike behavior from a monopolar (typically V+) DC power supply. However, not all op amp

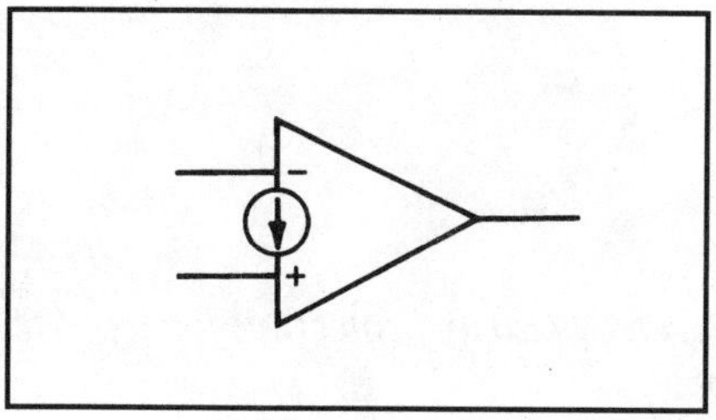

Figure 1-5. Symbol for current difference amplifier (CDA), also sometimes called a Norton amplifier.

performance will be available from some of these devices because the output voltage may not be able to assume negative values.

High Voltage. Most op amps operate at DC power supply potentials of ±6 to ±22 VDC. A few devices in the high voltage category operate from ±44 VDC supplies, and at least one proprietary hybrid model (not listed here) operates from ±100 VDC power supplies.

Multiple Devices. This category requires only that more than one op amp be included in the same package. Devices exist with two, three or four operational amplifiers in a single package. The 1458 device, for example, contains two 741-family devices in either 8-pin metal can or miniDIP packages. The 324 IC contains four single DC supply op amps.

Limited/Special Purpose Devices. The devices in this category are designed for either a limited range of uses, or highly specialized uses. The type 302 buffer, for example, is an op amp that is connected internally into the noninverting unity gain follower configuration. Some consumer audio devices also fit into this category. For example, there are several dual op amp devices used for stereo audio that are optimized for audio applications.

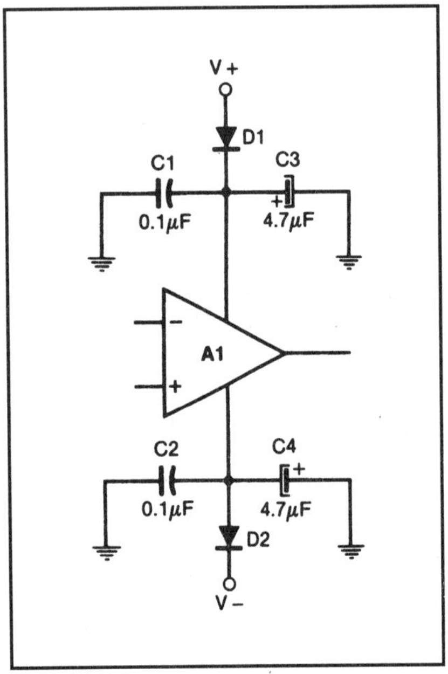

Figure 1-6. Reverse polarity protection for bipolar DC power supply.

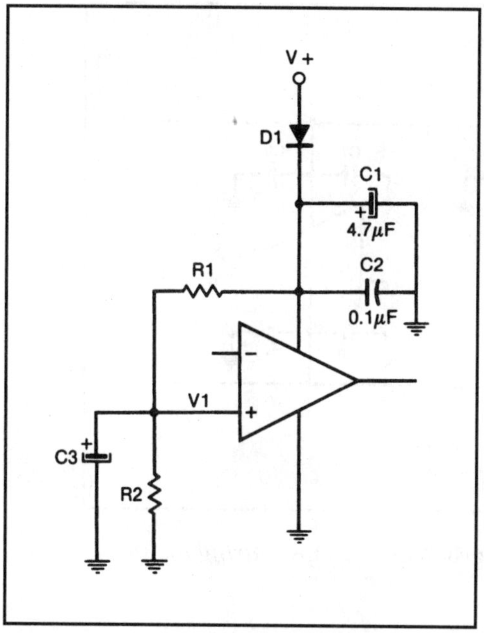

Figure 1-7. Reverse polarity protection for single polarity DC power supply.

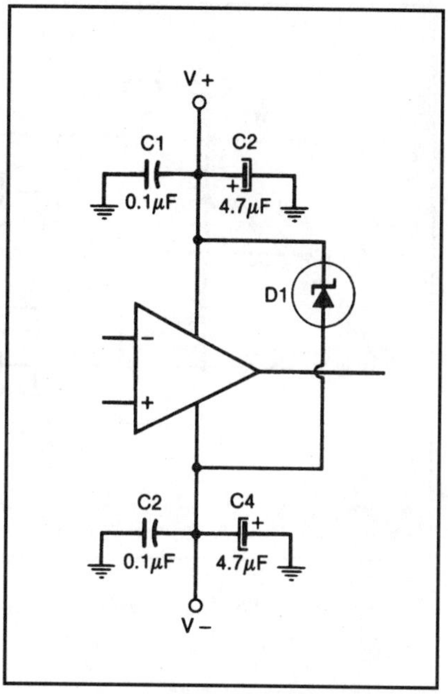

Figure 1-8. Reverse polarity protection using a zener diode between V- and V+.

Instrumentation Amplifiers. Although the instrumentation amplifier is arguably a "special purpose" device, it is sufficiently universal to warrant a class of its own. The "IA" is a DC differential amplifier made of either two or three internal op amps. Voltage gain can be set by either one or two external resistors. Chapter 7 deals with these amplifiers in detail.

Current Difference ("Norton") Amplifiers

Another form of IC linear amplifier is the Norton amplifier, also known as the current difference amplifier (or CDA). The CDA produces an output voltage that is proportional to the difference between two input currents. The operation of the CDA is not exactly analogous to the op amp (i.e., with the input voltages replaced by input currents), so will not be detailed here. Chapter 10 deals with CDA devices in detail.

The symbol for the CDA differs from the normal op amp symbol in order to distinguish its unique operation. The CDA symbol shown in Figure 1-5 is the

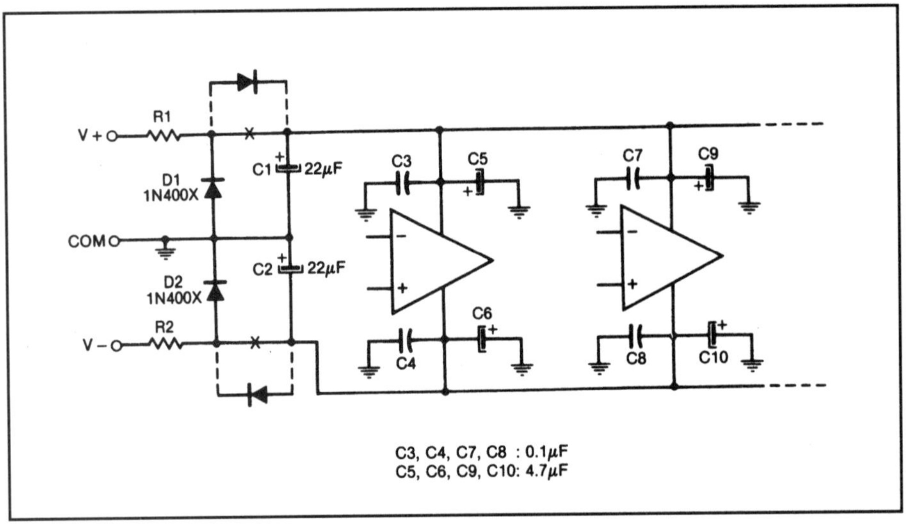

Figure 1-9. Reverse polarity protection for a group of op-amps sharing the same DC power supplies.

regular differential amplifier symbol with a current source symbol added along one edge to let the reader know that current mode is intended.

Operational Transconductance Amplifiers (OTA)

Another form of linear IC amplifier, different from either op amp or CDA, is the operational transconductance amplifier, or OTA. This type of amplifier has a transfer function that relates output current to input voltage. Since the transfer function expression has the units amperes/volts (or subunits thereof), the transfer function "gain" can be expressed in the units of conductance, siemens (or the sub-units millisiemens or microsiemens). Since these are units of conductance, I/V, we call the amplifier a transconductance amplifier. The name "operational" conveys the idea that some of the functions are similar to those of the operational amplifier.

Protecting Linear IC Amplifiers

Operational and other linear IC amplifiers are sensitive to problems on the DC power supply. The amplifier may oscillate if not properly decoupled. Further-

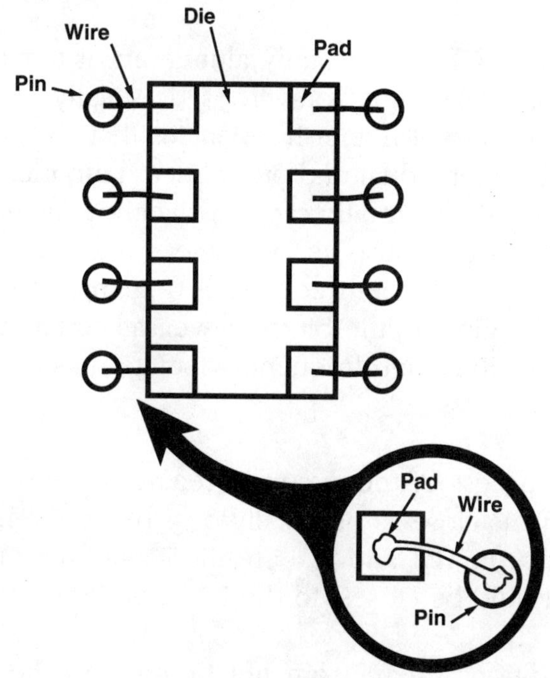

Figure 1-10. Integrated circuit die with wires connecting to package pins.

more, variations and noise placed on the DC power supply lines in one stage can affect the other stages — power supply rejection is not absolute.

We also find one other problem, especially in breadboarding and in portable (battery operated) equipment: reverse-polarity DC power supplies. The results may be catastrophic in this case! An operational amplifier with reversed DC power supplies can be destroyed instantly. Let's see how these problems can be fixed in practical circuits.

The problems with noise and oscillation are cured with decoupling capacitors on the amplifier power supply terminals. Capacitors C1 and C2 (each 0.1 μF) in Figure 1-6 are used to decouple high frequencies, while the low frequency decoupling is provided by C3 and C4 (each 4.7 μF or higher). Now why do you suppose that two forms of capacitors are needed at each op amp power supply terminal? That seems a bit odd, doesn't it? Why not just use the 4.7 μF capacitors — they are very much higher in value than the 0.1 μF units, after all.

The higher value capacitors (C3 and C4) are typically aluminum or tantalum electrolytics. The performance of these capacitors drops drastically as frequency increases, and may well be zero at higher frequencies that are nonetheless within the range of most IC operational and other linear amplifiers. At those elevated frequencies the typical electrolytic capacitor is about as effective as a block of wood. For this reason we also use a smaller value capacitor, but one that is of a type that will work at higher frequencies (e.g., mylar, mica, ceramic, etc.). This situation is changing a little bit, however, as certain new forms of capacitor are now able to offer high frequency operation as well as high capacitance.

One rule of thumb ensures the success of the circuit in regard to noise and oscillation: place those capacitors as close as physically possible to the body of the amplifier. The 0.1 μF capacitors (C1 and C2) are more important than the higher value capacitors, so should be closest to the IC amplifier body.

The fix for reverse-polarity conditions (apart from not doing it in the first place) is shown in Figure 1-6 also. Diodes D1 and D2 are placed in series with each DC power supply line. Under normal operation these diodes are forward-biased, so will conduct current to the amplifier. When some goofball like me accidentally connects one or both DC supplies backwards, then these diodes are reverse-biased so will not conduct current. Thus, the series diodes protect the amplifier IC from people's stupidity. Typical diodes for this application are any of the 1N400x series (1N4001 - 1N4007).

For single polarity DC power supplies we need only a single diode, as shown in Figure 1-7. Again, a 1N400x diode is both low cost and effective. In both cases (Figures 1-6 and 1-7), it is imperative that the capacitors be used as well as the diodes.

A brute force method of protecting the amplifier is shown in Figure 1-8. In this case, a zener diode is placed across the two DC power supply terminals of the amplifier. The zener potential must be greater than the maximum value of

[(V+) - (V-)]. For a case where DC power supplies of ±12 VDC are used, then this value is 24 volts. A 28-volt zener diode would be adequate, provided that the power supply voltages are reasonably stable. Under these conditions, with V_z greater than the voltage between the terminals, zener diode D1 is reverse-biased at a value lower than the zener potential. Thus, it is not used in normal operation.

In reverse-polarity operation, diode D1 becomes forward-biased in the non-zener mode. It will pass current around the amplifier harmlessly — hopefully. The big "maybe" is whether or not the zener diode will be destroyed. One solution is to use a very high value of power dissipation for D1. Another solution is to place a series resistor in the line with D1. All in all, it is better to use the alternative circuits using series diodes — but Figure 1-9 is provided for the truly bold reader.

The protection of multiple amplifier stages is shown in Figure 1-9. There are two alternatives shown in this figure. In one case, we could place 1N400x-series diodes across the power supply lines and series current-limiting resistors to prevent them from burning up. The diodes are normally reverse-biased, but when one or both DC power supplies are reversed, then these diodes become forward-biased — and short the line to ground.

The second alternative is to place the diodes in series with the line at the power supply terminals (shown in dotted lines in Figure 1-9). This method is analogous to the method of Figure 1-6, except that it serves more than one amplifier.

Common IC Package Types

The integrated circuit is formed on a tiny "chip" of silicon material by a photolithographic process. Typical chip "die" sizes range around 100-mils (0.100-inch), with some being larger and others being smaller. The die is typically mounted inside of a package and connected to the package pins by fine wires.

Figure 1-10 shows a die with wires attached. The connecting wires between package pins and "solder" pads on the die are around 10-mils (0.010-inch)

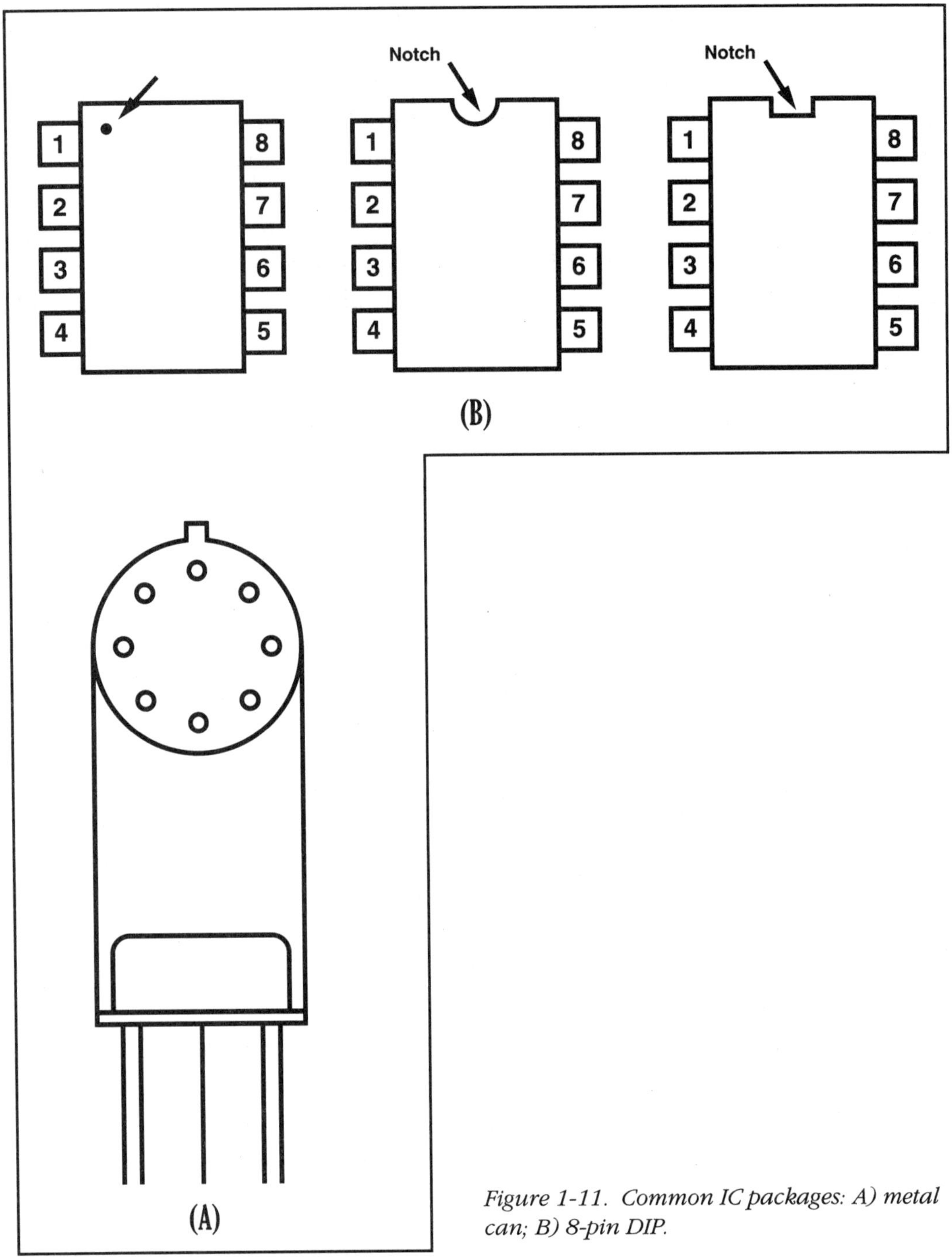

Figure 1-11. Common IC packages: A) metal can; B) 8-pin DIP.

diameter, and are made of either gold or aluminum in most cases. Either an electric current or a thermosonic process is used to melt the end of the wire onto (and bond it with) the connecting pad on the semiconductor die.

The particular package style selected for any given IC device depends in part on the intended application and the number of pins required. For many IC devices several different packages are available. The earliest IC packages were the 6, 8, 10 and 12-lead metal can devices (Figure 1-11A). These packages were redesigns of (and similar to) the TO-5 metal transistor package.

When viewed from the bottom the package, the keyway marks the highest number lead or "pin" (e.g., pin no. 8 in Figure 1-11A), and pin no. 1 is the next pin clockwise from the keyway. One must be careful when looking at IC base diagrams to know whether a top or bottom view is depicted.

Perhaps the largest number of IC devices on the market today are sold in dual in-line packages (DIP), examples of which are shown in Figure 1-11B. DIP packs are available in a wide variety of sizes from four to more than forty-eight pins. Although many devices are available in other size DIP packs, most linear devices are found in 8, 14 or 16-pin DIP packs.

The DIP pack is symmetrical with respect to pin count regardless of the package size, so some means is needed to designate pin no. 1; Figure 1-11B shows several common methods. In all cases the IC DIP pack is viewed from the top. On some devices a paint dot, "pimple," or "dimple" will mark pin no. 1. In other cases, a semicircular or square notch marks the end where pin no. 1 is located. When viewed from the top, with the notch pointed away from you, pin no. 1 is to the left of the notch while the highest number pin is to the right of the notch.

Both plastic and ceramic materials are used in DIP pack construction. In general, the plastic packages are used in consumer and noncritical commercial (or industrial) equipment, while the ceramic packs are used in military and critical commercial equipment. The principal difference between the plastic and ceramic packages is the specified temperature range. Although exceptions exist, typical temperature range specifications are 0 to +70 degrees Celsius for commercial plastic devices, and -10 to +80 degrees Celsius for ceramic. Military and some critical commercial ceramic devices are rated from -55 to +125 degrees Celsius. The principal difference between military chips and the highest grade commercial or industrial chips is the amount of testing, "burn-in" and documentation that accompanies each device.

Conclusion

We have discussed the operational amplifier, the current difference "Norton" amplifier, and the operational transconductance amplifier. We have also covered amplifiers in general, and methods of protecting IC amplifiers. Now let's get down to business, and find out how some of these circuits actually work. First, we will examine the ideal operational amplifier in Chapter 2, and then follow up with several chapters on operational amplifier applications.

The approach used in this book is the popular cookbook style. It was selected not only because it is popular, but also because it is easy to follow. Each case will be presented not just as a simple circuit that you could build (which you can if you want to), but also with the design steps laid out so that you can follow the logic and modify the circuit to fit your own needs. Please feel quite free to change and modify as it suits you.

Chapter 2

The Ideal Operational Amplifier

When you study any form of electronic device, be it vacuum tube, transistor or integrated circuit, it is wise to start with an ideal representation of that device and then proceed to less-than-ideal practical devices. In some cases, the practical and ideal devices are so far apart that you might wonder at the wisdom of this approach. In engineering and technical schools the students often take laboratory classes to put into practice what was learned in the theory classes. When they study their lab workbook sheets they often wonder if the theory instructor wasn't distributing a little saltwater taffy: the experiments often fail to come close to predicted theoretical behavior. However, for integrated circuit operational amplifiers, the modern commercial product so nearly approximates the ideal op amp of textbooks that the lab experiments actually work. The analysis method thus becomes extremely useful for learning the technology, designing new circuits, or doping out how someone else's circuit works.

In Chapters 3 and 4 we will discuss the inverting and noninverting operational amplifier configurations, respectively. In those discussions we will derive the design equations that you can use to make your own circuits — or understand mine. We will use a simplified analysis method that uses ordinary arithmetic rather than elegant engineering math. This method, while lacking the elegant mathematical rigor of an engineering textbook presentation, is nonetheless sufficiently powerful to permit you to understand and design other forms of circuits. The usefulness of our simplified approach proceeds directly from the correspondence of the ideal and practical operational amplifier IC devices.

Properties of the Ideal Op Amp

Our ideal op amp is characterized by several unique properties. From this short list of properties we can deduce circuit operation and design equations. Also, the list gives us a basis for examining non-ideal operational amplifiers and their defects (plus solutions to the problems caused by those defects). The basic properties of the op amp include:

1. Infinite open-loop voltage gain.
2. Infinite input impedance.
3. Zero output impedance.
4. Zero noise contribution.
5. Zero DC output offset.
6. Infinite bandwidth.
7. Both differential inputs stick together.

Let's take a look at these properties to determine what they mean in practical terms. You will find that real op amps only approximate these ideals; but in premium grade devices (and some lower cost units as well), the approximation is quite good.

Property No. 1: Infinite Open-Loop Gain. The "open-loop" gain of any amplifier (operational or not) is simply its gain without negative or positive feedback. Positive feedback is found in oscillators, but only in some amplifiers. For our purposes it is the negative feedback that is of concern. By definition, negative feedback is a signal fed back to the input 180 degrees out of phase. In operational amplifier terms this means feedback between the output and the inverting input.

Negative feedback has the effect of reducing the loop gain by a factor (called "β") that depends upon the transfer function and properties of the feedback network. Figure 2-1 shows the basic configuration for any negative feedback amplifier. The "transfer equation" for any circuit is the output function divided by the input function. For example, the transfer function of a voltage amplifier is V_o/V_{in}. In Figure 2-1 the term A_{vol} represents the gain of the amplifier element only; i.e., the gain with the feedback network disconnected. The term "β∀ represents the transfer function of the feedback network (without the amplifier). The overall transfer function of this circuit; i.e., with both amplifier element and feedback resistor in the loop, is defined as:

$$A_v = \frac{A_{vol}}{1-A_{vol}\beta} \qquad eq.\ (2\text{-}1)$$

— where:

A_v is the closed-loop gain of the circuit.
A_{vol} is the open-loop gain of the op amp.
β is the transfer equation for the feedback network.

In the ideal op amp A_{vol} is infinite, so the voltage gain is a function only of the feedback network. Remember this fact, for it is important enough to commit to memory very early in the game! In real op amps, the value of the open-loop gain is not infinite, but it is quite high. Typical values range from 20,000 in low-grade consumer models to more than 2,000,000 in premium units.

Property No. 2: Infinite Input Impedance. This property implies that the op amp input will not load the signal source. The input impedance of any amplifier is definable as the ratio of the input voltage and the input current: $Z_{in} = V_{in}/I_{in}$. When the input impedance is infinite, therefore, we must assume

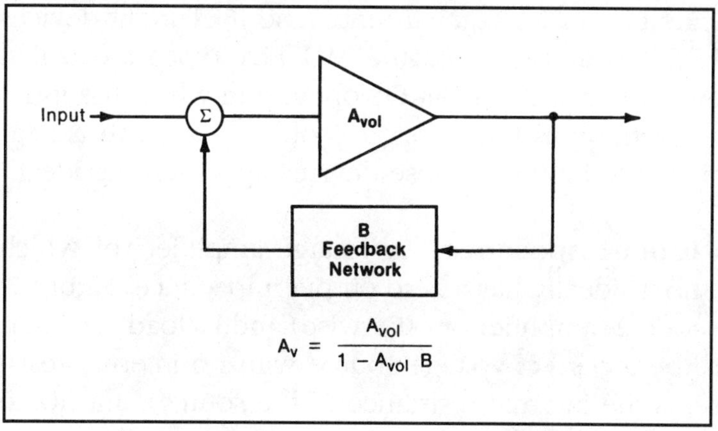

Figure 2-1. Feedback amplifier circuit equivalent circuit.

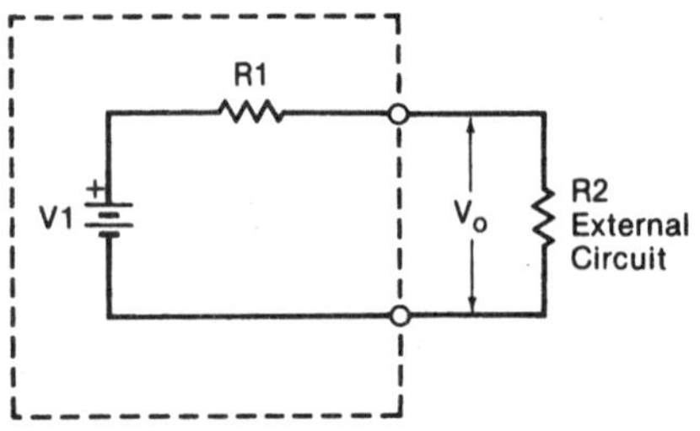

$$V_o = V\left(\frac{R2}{R1 + R2}\right)$$

Figure 2-2. Equivalent circuit for amplifier output circuit. Resistance R1 represents the internal output resistance of the device, while R2 represents the external load resistance.

that the input current is zero. Thus, an important implication of this property is that the operational amplifier inputs neither sink nor source current. In other words, it will neither supply current to a circuit, nor accept current from an external circuit. We will depend upon an implication of this property — $I_{in} = 0$ — to make our circuit analysis in Chapters 3 and 4.

Real operational amplifiers have some finite input current other than zero. In low-grade devices this current can be substantial (1 milliampere), and will cause a large output offset error. The primary source of this current is the base bias currents from the NPN and PNP transistors used in the input circuits. Certain premium-grade op amps that feature bipolar inputs reduce this current to scores of picoamperes (1 pA = 10^{-12} A). In op amps that use field effect transistors in the input circuits, on the other hand, the input impedance is quite high due to minuscule bias or leakage currents. The JFET input devices are typically called "BiFET" op amps, while the MOSFET types are called "BiMOS" devices. The CA-3140 device is a BiMOS op amp in which the input impedance approaches 1.5 terraohms (i.e., 1.5 x 10^{15} ohms) — which is near enough to "infinite" to make the inputs of those devices approach the ideal.

Property No. 3: Zero Input Impedance. A voltage amplifier (of which class the op amp is a member) ideally has a zero output impedance. Figure 2-2 represents any voltage source (amplifier or otherwise) and its load, or external circuit. Consider V to be a perfect voltage source with no internal resistance (resistor R1 represents the internal resistance of the source), and R2 to be the load. Because the internal resistance (which in amplifiers is usually called "output resistance") is in series with the load resistance, the output voltage V_o that is available to the load is reduced by the voltage drop across R1. Thus, the output voltage is:

$$V_o = \frac{VR2}{R1+R2} \qquad \text{eq. (2-2)}$$

It is clear from the above equation that the output voltage will equal the source voltage only when the output resistance of the amplifier (R1) is zero. In that case, Vo = V x (R2/R2) = V. Thus, in the ideal voltage source, or the ideal op amp, we get the maximum output voltage (and the least error) because no voltage is dropped across the internal resistance of the amplifier.

Real operational amplifiers do not have a zero output impedance. The actual value is typically less than 100 ohms, with many being in the neighborhood of 50 ohms. Thus, for typical voltage amplifier inputs driven by the operational amplifier, the output can be taken as if it were ideal.

A rule of thumb used by designers is to make the input resistance of any device that is driven by a non-ideal output at least ten times the device output impedance. We see this situation in Figure 2-3. Amplifier A1 is a voltage source that drives A2. In this situation, R1 represents the output resistance of A1 and R2 represents the input resistance of amplifier A2. In practical terms, we find that the circuit where R2 > 10R1 will yield results acceptably close to "ideal" for most purposes. In some cases, the rule R2 > 100R1 must be followed, but these are rare and are only those that require precision.

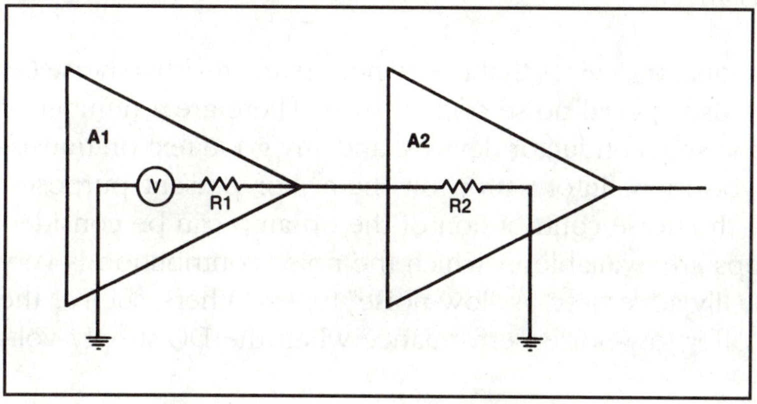

Figure 2-3. Equivalent circuit for cascade amplifiers.

Property No. 4: Zero Noise Contribution. All electronic circuits, even simple resistor networks, produce noise signals. An example of electronic noise can be seen by shorting together the antenna input terminals of your FM receiver and then turning the volume control all the way up. The hiss you hear is the noise that is generated inside the electronic circuitry of the receiver. If the noise increases when the antenna is connected, then the increase represents atmospheric or natural noise.

In the ideal operational amplifier, all noise voltages produced internally amount to zero. Thus, any noise in the output signal must have been in the input signal as well. Except for amplification, the output noise voltage will be exactly the same as the input noise voltage. In other words, the op amp contributed nothing extra to the output noise.

Practical op amps do not approximate the ideal, except for certain high-cost models. Even a simple resistor creates noise due to the movement of electrons on its insides. For all resistive circuits, the noise contribution is:

$$V_{noise} = \sqrt{4KTBR} \qquad eq. (2\text{-}3)$$

— where:

K is Boltzman's constant (1.38 X 10^{-16} erg/(K).
T is the temperature in degrees Kelvin (°K).
B is the bandwidth in hertz (Hz).
R is the resistance in ohms (Ω).

As you can easily see, the noise voltage produced by a resistor varies with the square root of its resistance.

Amplifiers use semiconductor devices that create not merely resistive noise (as described above), but also special noise of their own. There are a number of internal noise sources in semiconductor devices, and any good text on transistor theory will give you more information on them. For present purposes, however, assume that the noise contribution of the op amp can be considerable. Premium op amps are available in which the noise contribution is very low, and these are usually advertised as "low-noise" types. Others, such as the CA-3140 device, will offer low-noise performance when the DC supply voltages are limited to ±5 volts, and the metal package of the op amp is fitted with a flexible "TO-5" style heat sink.

Property No. 5: Zero Output Offset. The output offset voltage of any amplifier is the output voltage that exists when it should be zero. The voltage amplifier sees a zero input voltage when the inputs are both grounded. This connection should produce a zero output voltage. If this voltage is non-zero, then there is said to be an output offset voltage present. In the ideal op amp, this offset voltage is zero volts — but real op amps aren't so lucky.

In the op amps the output offset voltage is non-zero, although it can be quite low in some cases. There are several methods for dealing with output offset voltage, and these will be discussed at length in Chapter 5. Most readers of this book are electronic hobbyists or students, and often buy components from certain "bargain" sources. Be aware that many of the "bunches of op amps for less than a buck" deals are possible only because the op amps are factory seconds (bought wholesale for pennies per shovelful). In one famous chain's offering, I found that six of the ten 741 op amps in a single package had more than 1 volt of output offset potential. Always test those cheapie op amps (or other IC chips) before using them — that way you'll be able to tell a bad circuit or construction error from a bad op amp device.

Property No. 6: Infinite Bandwidth. The ideal op amp will amplify all signals from DC to daylight — clearly a ridiculous claim for real op amps. In real op amps, the bandwidth is sharply limited. There is a specification called the "gain-bandwidth product" (GBP), which is symbolized by F_t. This rating is simply the frequency at which the gain drops to unity (1). The maximum available gain is found from dividing the maximum required frequency into the gain-bandwidth product. If the value of F_t is not sufficiently high, then the circuit will not behave in classical op amp fashion at some frequencies.

Some practical op amps have GBPs in the 2 to 10 MHz range. Others, on the other hand, are quite limited. The 741-family of devices are very limited, such that the device will perform as an op amp for only a few kilohertz. Above that range, the gain drops off considerably. However, in return for this apparent limitation, we obtain unconditional stability. Such op amps are said to be "frequency compensated." It is the frequency compensation of those devices that kills off the GBP. Non-compensated op amps will yield wider frequency response, but at the expense of certain instabilities. Those op amps may oscillate at some frequency without any special encouragement (see Chapter 5).

Property No. 7: Both Differential Inputs Stick Together. Although differential inputs are not part of the definition of operational amplifiers, and indeed at least one major manufacturer offers op amps with only an inverting input, most op amps have two inputs: one inverting (-) and the other noninverting (+).

By "stick together," I mean that a voltage applied to one of these inputs also appears at the other input. This voltage is real, and is not merely some theoretical device used by engineers to evaluate circuits. If you apply a voltage to, say, the inverting input, and then connect a voltmeter between the noninverting input and the power supply common (or ground), then the voltmeter will read the same potential on noninverting as on the inverting input.

The implication of this property is that we must treat both inputs the same mathematically. This fact will make itself felt in Chapter 3 when we discuss the needlessly confusing concept of "virtual" as opposed to actual grounds, and again in Chapter 4 when we deal with the noninverting follower circuit configuration.

Conclusion

Our discussion thus far has not dealt with the problems caused by the departures from the ideal properties offered by real, practical "kind you can actually buy" operational amplifiers. For that discussion you may turn to Chapter 5. In the meantime, however, we will want to first develop the two major configurations of the operational amplifier: inverting follower (Chapter 3) and noninverting follower (Chapter 4).

The inverting follower produces an output signal that is 180 degrees out of phase with its input signal. These circuits are covered in detail in Chapter 3. The noninverting follower, as you might expect, produces an output signal that is in-phase with its input signal. These circuits are covered in Chapter 4. Almost all other operational amplifier circuits are variations on either the inverting or noninverting follower. Understanding these two configurations will make you able to understand, and design or modify, a wide variety of different circuits.

Chapter 3

Inverting Follower Op Amp Circuits

The inverting follower is an op amp configuration in which the output signal is 180 degrees out of phase with the input signal. To accomplish this neat trick we use the inverting ("-") input of the operational amplifier.

Figure 3-1 shows the basic configuration for inverting follower (also called inverting amplifier) circuits. The noninverting input is not used, so is set to ground potential. There are two resistors in this circuit. Resistor R2 is the negative feedback path from the output to the inverting input, while R1 is the input resistor. In a moment we will examine R1/R2 in order to determine how gain is fixed in this type of circuit. First, let's take a look at the implications of grounding the noninverting input in this type of circuit.

What's a "Virtual" Ground?

The concept of "virtual ground" sounds like a ground that only pretends to be a ground, and in a funny kind of way that's not an unreasonable description. Unfortunately, that type of terminology is both confusing and leads to an implication that the virtual ground doesn't really function as a ground. Let's clear up that little mess, okay?

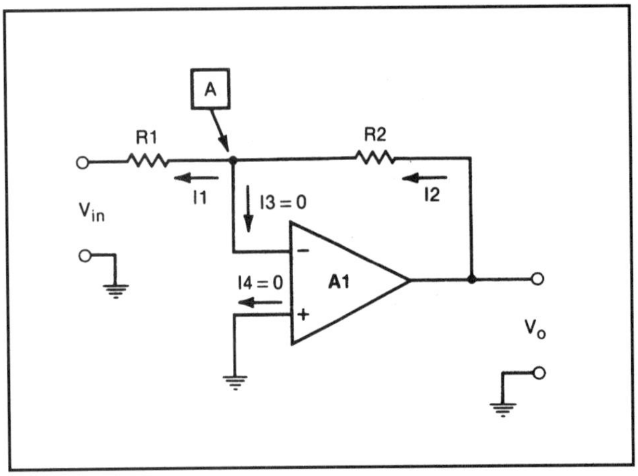

Figure 3-1. Inverting follower circuit.

In Chapter 2 you learned the properties of the "ideal operational amplifier." One of those properties tells us that differential inputs "stick together." Put another way, this property means that a voltage applied to one input appears on the other input also. In the arithmetic of op amps, therefore, we must treat both inputs as if they were at the same potential. This fact is not merely a theoretical device, either, for if you actually apply a potential, say 1 volt DC, to the noninverting input the same 1 volt DC potential will also be measured with a real voltmeter at the inverting input (try it).

In Figure 3-1, the noninverting input is grounded, so is at zero volts potential. This fact, by the properties of the ideal op amp, means that the inverting input of the op amp is also at ground potential (zero volts). Since the inverting input is at ground potential, but has no physical ground connection, it is said to be at "virtual" (as opposed to physical) ground.

The choice of the term "virtual ground" was unfortunate, for the idea of a virtual ground is actually quite simple even though the terminology makes it sound a lot more abstract.

Developing the "Transfer Equation" For the Inverting Follower

The "transfer equation" of a circuit is merely the output function divided by the input function. For an operational amplifier used as a voltage amplifier, the transfer function describes the gain, and is:

- $A = \frac{V_o}{V_{in}}$ *eq. (3-1)*

— Where:

A is the voltage gain (dimensionless).
V_o is the output signal potential.
V_{in} is the input signal potential (V_o and V_{in} in the same units)

Example

What is the voltage gain if 100 mV (0.100 volt) at the input produces a 10 volt output potential?

$A = V_o/V_{in}$
$A = 10/0.100 = 100$

In the op amp inverting follower circuit (Figure 3-1) we find that the gain is set by the two resistors, R1 and R2. Let's do a step-by-step analysis to see if we can figure out the relationship. The method we use is simplified, but is valid for most op-amp circuits so you might want to pay as much attention to how it's done as to the result.

Consider the currents flowing in Figure 3-1. The input bias currents, I3 and I4, are assumed to be zero. In a real op amp they are non-zero, and have to be accounted for, but in the analysis case we use ideal "textbook" op amps rather than the kind you can buy on the hobbyist and amateur markets. Thus, in the analysis below we can ignore I3 = I4 = 0.

Remember that the summing junction (point "A") is at ground potential because the noninverting input is grounded. Because of this fact, current I1 is a function of only the applied input voltage, V_{in}, and the resistance R1. By Ohm's law, then, the value of I1 is:

- $I1 = \frac{V_{in}}{R1}$ *eq. (3-2)*

Further, we know that current I2 is also related by Ohm's law to the output voltage, V_o, and the feedback resistor R2 (again, because the summing junction is grounded):

- $$I2 = \frac{V_o}{R2}$$ eq. (3-3)

How are I1 and I2 related? These two currents are the only currents entering or leaving the summing junction (recall that I3 = 0), so by Kirchoff's Current Law (KCL) we can assume:

- $$I2 = -I1$$ eq. (3-4)

We begin to home in on the transfer function by substituting eqs. (3-2) and (3-3) into eq. (3-4):

- $$\frac{V_o}{V_{in}} = \frac{-V_{in}}{R1}$$ eq. (3-5)

Rearranging eq. (3-5) yields the formal version of the transfer equation:

- $$\frac{V_o}{V_{in}} = \frac{-R2}{R1}$$ eq. (3-6)

According to eq. (3-1): the gain (A) is V_o/V_{in}, so we may write eq. (3-6) in the form:

- $$A_v = \frac{-R2}{R1}$$ eq. (3-7)

Example

What is the gain of an op amp inverting follower if the input resistor (R1) is 10K, and the feedback resistor is 1 megohm?

A = -R2/R1
A = -(1,000,000)/10,000 = -100

Thus, we have shown that the voltage gain of an op amp inverting follower is merely the ratio of the feedback to the input resistor (-R2/R1). The minus sign indicates that a 180 degree phase reversal takes place. Thus, a negative input voltage produces a positive output voltage, and vice versa.

We often see the more practical form of transfer eq. (3-6), written to express output voltage in terms of gain and input signal voltage. The two expressions are:

- $V_o = -A_v V_{in}$ *eq. (3-8)*

— and,

- $V_o = -V_{in}\frac{R2}{R1}$ *eq. (3-9)*

Example

What is the output voltage from an inverting follower in which V_{in} = 100 mV (0.100 V), R2 = 100K and R1 = 2K?

V_o = -Vin(R2/R1)
V_o = -(0.100 volts)(100K/2K)
V_o = -(0.100 volts)(50) = -5 volts

In the example above, the voltage gain (A) is R2/R1 = 100K/2K = 50.

The following equations are used for inverting followers, and should be memorized:

- $A = \frac{-R2}{R1}$ *eq. (3-10)*
- $V_o = -AV_{in}$ *eq. (3-11)*
- $V_o = -V_{in}\frac{R2}{R1}$ *eq. (3-12)*

Frequency Response Tailoring

We can tailor the upper-end frequency response of the operational amplifier with a capacitor shunting the feedback resistor (see Figure 3-2). This capacitor, acting on the value of R2, sets the frequency at which the upper-end frequency response falls off to -3 dB below the low-end gain. The gain at fre-

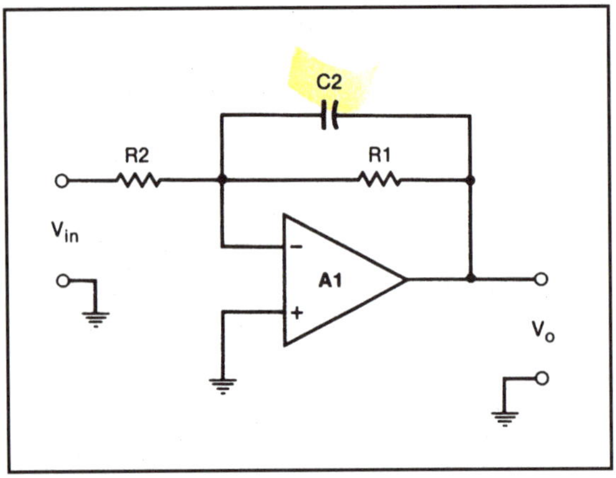

Figure 3-2. Capacitor rolls off high frequency response.

quencies higher than this -3 dB frequency falls off at a rate of -6 dB/octave (an octave is a 2:1 frequency ratio), or -20 dB/decade (a decade is a 10:1 frequency ratio).

The value of capacitor C2 is found by:

$$C2 = \frac{1{,}000{,}000}{2\pi R2F} \qquad \text{eq. (3-13)}$$

— where:

C2 is the capacitance in microfarads (µF).
R2 is in ohms (Ω).
F is the upper -3 dB frequency in hertz (Hz).

Example

Find the value of capacitance required to shunt across R2 in Figure 3-2 in order to achieve a -3 dB frequency of 3,000 Hz when the feedback resistor is 470K.

$C2 = 1{,}000{,}000/(2\pi R_2F)$
$C2 = 1{,}000{,}000/[(2)(3.14)(470{,}000 \text{ ohms})(3{,}000 \text{ Hz})]$
$C2 = 1{,}000{,}000/8.85 \times 10^9$
$C2 = 0.0001\ \mu F = 100 \text{ pF}$

The low frequency response is controllable by placing a capacitor in series

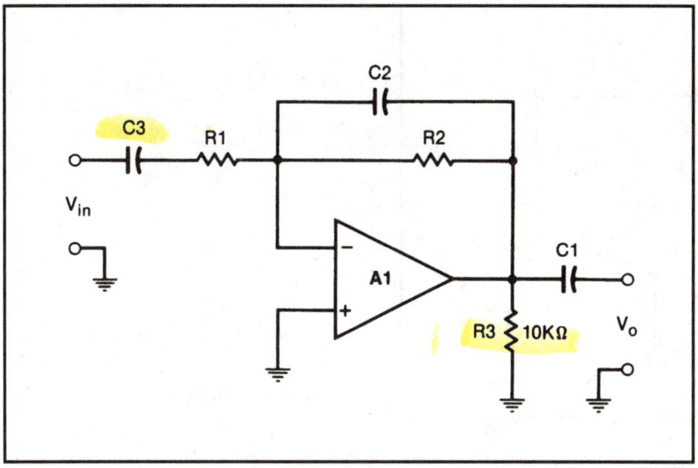

Figure 3-3. AC-coupled inverting follower.

with the input resistor, which also makes the inverting follower an AC-coupled amplifier.

AC-Coupled Inverting Followers

Figure 3-3 shows the circuit for an AC-coupled inverting follower. Capacitor C2 limits the upper -3 dB frequency response point. Its value is set by the method discussed above. The lower -3 dB point is set by the combination of R1 and input capacitor C1. This frequency is set by the equation:

$$C3 = \frac{1{,}000{,}000}{2\pi R1F} \qquad \text{eq. (3-14)}$$

— where:

C3 is in microfarads (μF).
R1 is in ohms (Ω).
F is the lower -3 dB point in hertz (Hz)

The circuit of Figure 3-3 is an AC-coupled amplifier. In some cases we will want to also AC-couple the output circuit (although this is optional in most cases). Capacitor C1 is used to AC-couple the output, thus preventing any DC component on the op amp output from affecting the following stages. Resistor R3 is used to keep capacitor C1 from being charged by the offset voltage from

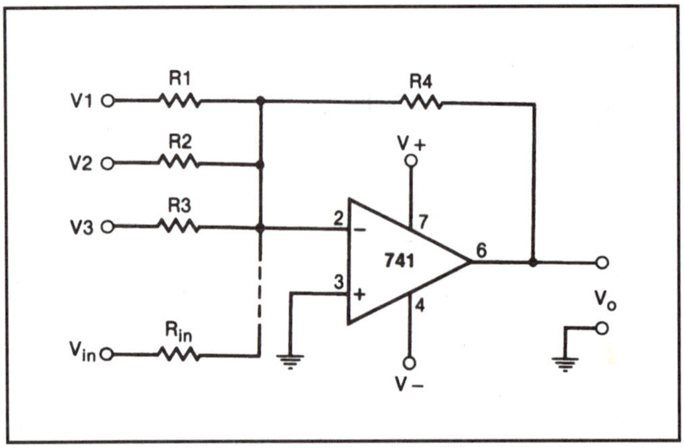

Figure 3-4. Multiple input inverting follower (also called a summing amplifier).

op amp A1. The value of capacitor C1 is set to retain the lower -3 dB point, using the resistance of the stage following as the "R" in the equations above.

Multiple-Input Inverting Followers

We can accommodate multiple inputs on an inverting follower by using a circuit such as Figure 3-4. There are a number of applications of such circuits: summers, audio mixers and so forth (for example). The output voltage transfer function for this type of amplifier (eq. [3-15]) is listed under the following example. The terms "V_n" and "Rn" refer to the nth voltage and nth resistance. Let's consider a three-input example.

Example

A circuit such as Figure 3-4 has a 100 kohm feedback resistor (R4), and the following input resistors: R1 = 10K, R2 = 50K, and R3 = 100k. Find the output voltage when the following input voltages are present: V1 = 0.100 volts, V2 = 0.200 volts and V3 = 1 volt.

- $$V_o = R4\left(\frac{V1}{R1} + \frac{V2}{R2} + \frac{V3}{R3} + \ldots + \frac{V_n}{R_n}\right) \qquad \text{eq. (3-15)}$$

$$V_o = (100K)\left(\frac{0.1V}{10K} + \frac{0.2V}{50K} + \frac{1V}{100K}\right)$$

$$V_o = (10)(0.1) + (2)(0.2) + (1)(1.00)$$

$$V_o = (1V) + (0.4V) + (1V) = 2.4\ volts$$

Basic Design Procedure

We must consider several matters when designing inverting follower amplifiers. First, we must obviously consider the gain required — that's basic. Second, we must consider the input impedance of the circuit. That specification is needed in order to prevent the amplifier input from loading the driving circuit. In the case of the inverting follower, the input impedance is limited to the value of the input resistor (R1 in Figure 3-1). We must follow the design rule:

RULE No. 1: MAKE THE INPUT RESISTOR (HENCE THE INPUT IMPEDANCE) NOT LESS THAN 10 TIMES THE SOURCE RESISTANCE (MORE IS BETTER).

The implication of this rule is that we must determine the source resistance of the driving circuit, and then make the input impedance of the operational amplifier inverting follower at least ten times larger. When the driving source is another operational amplifier we can assume that the source impedance (i.e., the output impedance of the driving op amp) is 100 ohms or less. For these cases, make the value of R1 (in Figure 3-1) at least 1000 ohms (i.e., 10 x 100 ohms = 1000 ohms). In other cases, however, we have a slightly different problem. Some transducers — a thermistor for measuring temperature for example — have a much higher source resistance. One thermistor has a resistance that varies from 10K to 100k over the temperature range, so an input impedance of 1 megohm is required. When the input impedance gets this high we might want to consider the noninverting follower amplifier (see Chapter 4).

In the inverting follower circuit, the choice of input impedance drives the design, so part of the initial procedure is:

> Determine the minimum allowable input resistance
> (i.e., 10X the source impedance).

If the source resistance is 100 ohms or less, try 10 K as the trial input resistance (R1). This value might be lowered if the feedback resistor becomes too high for the required gain. The value of R1 is the input resistance, or 10K, whichever is higher.

Determine the amount of gain required. In general, the gain of a single inverting follower should be less than 500. For gains higher than that figure use a multiple op amp circuit. Some low-cost op amps should not be operated at gains greater than 200.

Determine the frequency response (i.e., the frequency at which the gain drops to unity). From steps 3 and 4 we can calculate the minimum gain bandwidth product of the op amp required. (G-B = gain x frequency)

Select the operational amplifier. If the gain is high, e.g., over 100, then you might want to select a BiMOS or BiFET operational amplifier in order to limit the output offset voltage caused by the input bias currents. Select a 741-family device if (a) you don't need more than a few kilohertz frequency response, and (b) the unconditionally stable characteristics of the 741 is valuable. Also look at the package style. For most hobbyist applications, the 8-pin mini-DIP package is probably the easiest to handle. The 8-pin metal can is also useful because it can be made to fit 8-pin mini-DIP positions by correct bending of the leads.

Select the value of the feedback resistor:

- $$R2 = |A|R1 \qquad \text{eq. (3-16)}$$

Note: "|A|" means the "absolute value of gain" — i.e., the gain without the minus sign.

If the value of the feedback resistor is too high — i.e., beyond the range of standard values (about 20 megohms or so) — or too high for the input bias currents, then try a lower input resistance (in the case where the value of R1 is greater than 10 times the source resistance).

If a high-end frequency response limit is imposed, then calculate the value of the capacitor shunting the feedback resistor (Figure 3-2) according to the equation:

$$C2 = \frac{1{,}000{,}000}{2\pi R2F} \qquad eq.\ (3\text{-}17)$$

— where:

C2 is the capacitance in microfarads (μF).
R2 is the resistance in ohms (Ω).
F is the upper -3 dB frequency response break point in hertz (Hz).

Project Examples

Since building circuits and projects is one of the best ways to learn electronics, the following examples are provided.

Case No. 3-1: Gain-of-Ten Preamplifier

Design an inverting amplifier that will produce an amplification of 10X for a DC potential being applied to a meter input in order to improve resolution (the meter's scale is 0 - 10 volts). Assume that the source impedance is 50 ohms, and that the input voltage is not greater than one volt.

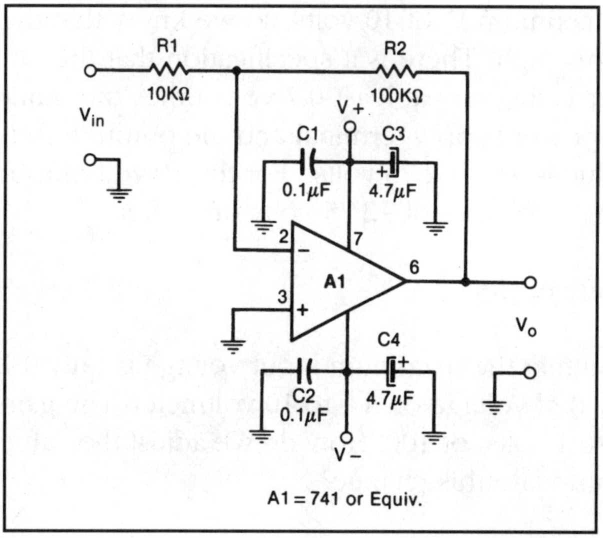

Figure 3-5. A gain-of-10 inverting amplifier.

Discussion

The low source impedance means that any input resistance greater than 500 ohms is sufficient. Given the moderate gain requirement, therefore, we can set the input resistor (R1) value at 10 K. Since the device operates at DC and near-DC, and is relatively healthy (one the order of 1 volt), we can use any ordinary 741-family operational amplifier. The circuit for this project, including pin-outs for the "industry standard" 741 operational amplifier, is shown in Figure 3-5.

The feedback resistor, R2, is selected from the expression:

- $$R2 = |A|R1 \qquad \text{eq. (3-18)}$$

$$R2 = |-10|x10K$$

$$R2 = (10)(10K) = 100K$$

The power supply pins for the 741 operational amplifier are as follows:

- v+ is pin 7 *eq. (3-19)*
  v- is pin 4

The values for V- and V+ are dependent on the maximum required output voltage. In this case, we have a maximum V_o of 10 volts, so we know that the DC supply voltages are at least this high. There is a specification that the DC voltage be higher than the output voltage by at least 0.7 volts times the number of PN junctions between the power supply terminal and the output terminal. In the case of the 741 this value is about 2.75 volts. For the 10 volt output case, therefore, we need V- and V+ voltages of 12.75 volts or higher.

Case No. 3-2: Gain of 100 Preamplifier

Assume in the case above we find that the maximum input voltage is only 0.1 volts, and we still need to display that voltage on a 0 to 10 voltmeter. The gain of this amplifier must be 10 volts/0.1 volts, or 100. How do we adjust the value of feedback resistor R2 to accommodate this change?

- $$R2 = |(-100)|(10K) \qquad \text{eq. (3-20)}$$

$$R2 = (100)(10K) = 1{,}000{,}000 \; ohms$$

The circuit for this amplifier is shown in Figure 3-6. The DC power supply connections are as shown in Figure 3-5.

Case No. 3-3: Gain-of-100, 100 Hz Inverting Amplifier With High Input Impedance

Suppose we have a source that produces 0.010 volts against a requirement of -1 volts input to an AC amplifier. This source is a temperature transducer with an output impedance of 10K or less. Because of the rapid rise-time of the transducer we need a 100 Hz frequency response, but wish to limit the higher frequencies to reduce noise in the circuit (some temperature transducers are inherently noisy). Design an amplifier for this application.

Discussion

The 10K maximum value of the transducer impedance means that we need to set R1 at 100K. The gain required is (-1)/(0.10) = -100, which implies a gain-of-100 inverting amplifier. Because of the high impedances involved, one needs to consider a BiMOS or BiFET operational amplifier. Because the RCA CA-3140 device is easily available to hobbyists, it would make a good choice.

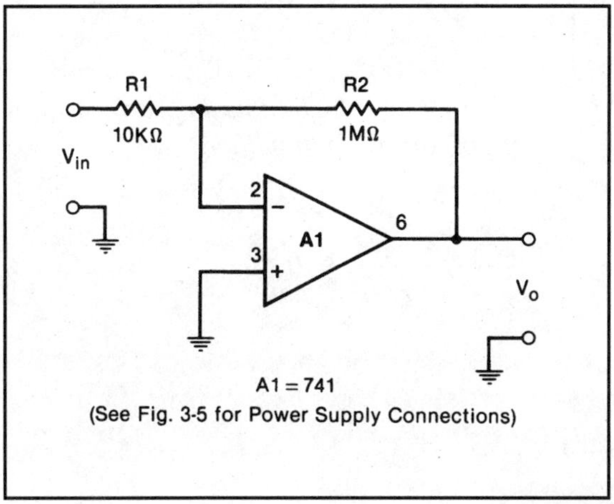

Figure 3-6. A gain-of-100 inverting amplifier.

Capacitor C1 limits the upper frequency of operation by setting the -3 dB point in the frequency response curve:

- $$C1 = \frac{1{,}000{,}000}{2\pi R1F}$$ *eq. (3-21)*
- $$C1 = \frac{1{,}000{,}000}{(2)(3.14)\left(10^{7} ohms\right)(100Hz)}$$ *eq. (3-22)*
- $$C1 = 0.000159\mu F = 159pF$$ *eq. (3-23)*

Because 160 pF is the nearest standard value, we select 160 pF for Figure 3-7.

Conclusion

In this chapter we discussed the inverting follower configuration. In the next chapter we will expand our discussion of basic op amp circuits by looking at the noninverting follower configuration.

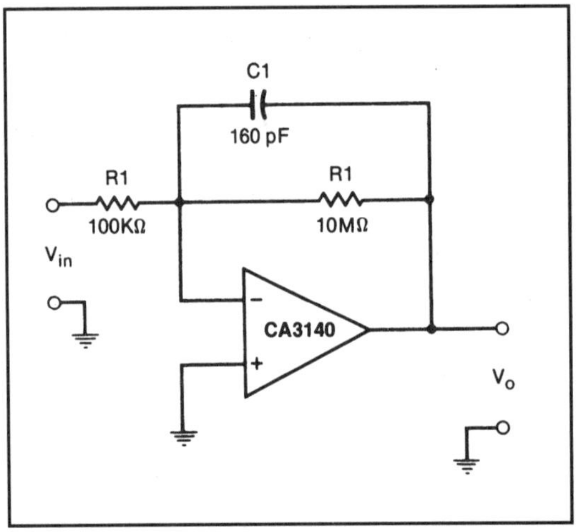

Figure 3-7. A high input impedance gain-of-100 inverting amplifier.

Chapter 4

Noninverting Followers

The second classical op amp configuration is called the noninverting follower. This type of amplifier uses the noninverting input of the operational amplifier to receive a signal. Because of this configuration, the output signal is in phase with the input signal. There are two basic noninverting configurations: unity gain and greater-than-unity gain.

Figure 4-1 shows the unity gain noninverting follower circuit. The output terminal is connected directly to the inverting input, resulting in 100 percent negative feedback. Recall the voltage gain expression for all feedback amplifiers:

$$A_v = \frac{A_{vol}}{1+A_{vol}\beta} \qquad eq.\ (4\text{-}1)$$

— where:

A_v is the closed-loop voltage gain (i.e., gain with feedback).
A_{vol} is the open-loop voltage gain (i.e., gain without feedback).
β is the feedback factor

The feedback factor, β, represents the transfer function of the feedback network. When that network is a resistor voltage divider network, the value of β is a decimal fraction that represents the attenuation of the op amp output voltage before it is applied to the op amp inverting input. In the unity gain follower circuit the value of β is 1; so eq. (4-1) reduces to eq. (4-2). Consider the implications of eq. (4-2) for a common op amp. A certain low-cost type has an open-loop gain of 50,000. For this device, the voltage gain of a circuit such as Figure 4-1 is:

$$A_v = \frac{A_{vol}}{1+A_{vol}} \qquad \textit{eq. (4-2)}$$

$$A_v = \frac{50{,}000}{1+50{,}000}$$

$$A_v = \frac{50{,}000}{50{,}001} = 0.99998$$

A gain of 0.99998 is close enough to 1.0 to justify calling the circuit of Figure 4-1 a "unity" gain follower.

Application of Unity Gain Followers

So what use is an amplifier that does not amplify? There are two principal uses of the unity gain noninverting follower: buffering and impedance transformation. A secondary use is boosting the drive power available (a special case of buffering).

A "buffer amplifier" is placed between a circuit and its load in order to improve the isolation of the two. A very common example is use of a buffer

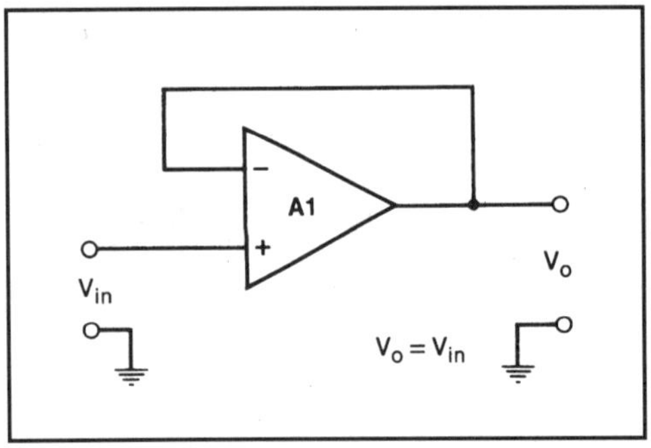

Figure 4-1. Unity gain noninverting follower.

amplifier between an oscillator or waveform generator and its load. This use is found especially where the load exhibits a varying impedance that could result in pulling of the oscillator frequency. Such unintentional frequency modulation of the oscillator is very annoying because it makes some circuits unable to function and others function poorly.

Another common use for buffer amplifiers is isolation of an output connection from the main circuitry of an instrument. An example might be an instrumentation circuit that uses multiple outputs, perhaps one to a computer and another to an oscilloscope or strip chart recorder. By buffering the analog output to the 'scope we prevent short circuits in the display wiring from affecting the signal to the computer.

A special case of buffering is represented by using the unity gain follower as a power driver. A long cable run will attenuate low-power signals. To overcome this problem we sometimes use a low-impedance power source to drive the long cable. This application points out that a "unity" gain follower does have power gain (the "unity" feature refers only to the voltage gain). After all, if the input impedance is typically much higher than the output impedance (yet $V_o = V_{in)}$, then it stands to reason that the power output is much greater than the required input power. Thus, the circuit of Figure 4-1 is unity gain for voltage signals and greater-than-unity gain for power.

The impedance transformation capability is obtained from the fact that an op amp has a very high input impedance and a very low output impedance. Let's illustrate this application by a practical example. Figure 4-2A is a generic equivalent of a voltage source driving a load (R2). The resistance R1 represents the internal impedance of the signal source (usually called "source impedance").

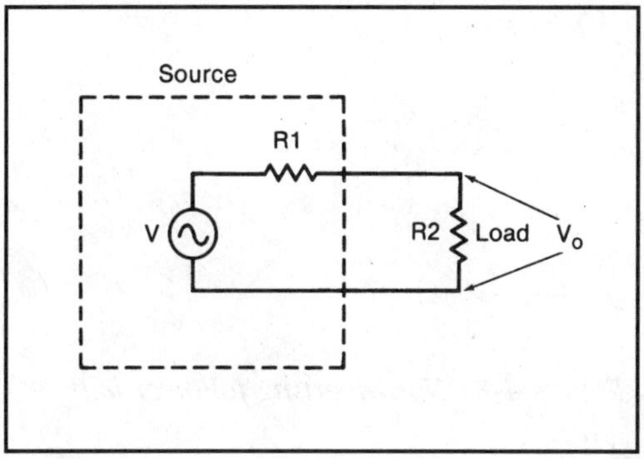

Figure 4-2. Generic equivalent of voltage source driving a resistive load.

The pure signal voltage, V, is reduced at the output (V_o) by whatever voltage is dropped across source resistance R1. In our "worst case" example let's assume a high impedance source is driving a low impedance load. We will calculate the loss in a case where R1 = 10K, R2 = 1K, and V = 1 volt.

$$A_v = \frac{50{,}000}{1+50{,}000} \qquad \text{eq. (4-3)}$$

$$A_v = \frac{50{,}000}{50{,}001} = 0.99998$$

$$V_o = \frac{VR2}{R1+R2}$$

$$V_o = \frac{(1V)(1K)}{(10K+1K)}$$

$$V_o = \frac{1}{11}\ volts = 0.091\ volts$$

We have lost more than 90 percent of the signal voltage. However, if we impedance transform the circuit with a unity gain noninverting amplifier, as in Figure 4-2B, we change the situation entirely. If the amplifier input impedance is very much larger than the source resistance (it is), and the amplifier output impedance is very much lower than the load impedance, then there is very little loss and V will closely approximate V_o.

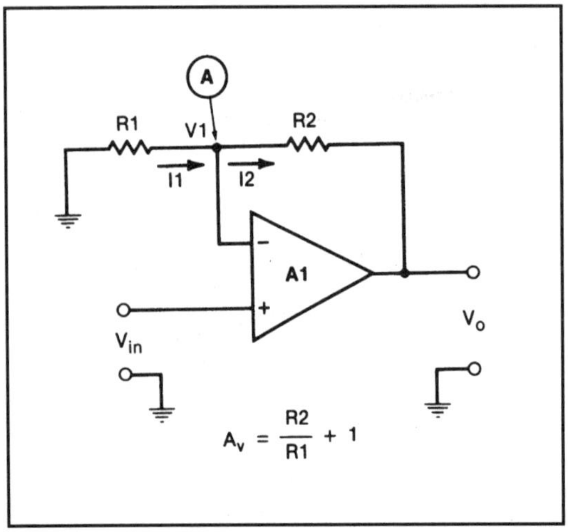

Figure 4-3. Noninverting follower with gain.

Noninverting Followers With Gain

Figure 4-3 shows the circuit for the noninverting follower with gain. In this circuit, the signal, V_{in}, is applied to the noninverting input, while the feedback network (R2/R1) is almost the same as in the inverting follower circuit.

We can evaluate this circuit using the same general method as was used in the inverting follower case (see Chapter 3). We know from Kirchoff's Current Law (good ol' KCL), and the fact that the op amp input neither sinks nor sources current, that I1 and I2 are equal to each other. Thus, for these currents at the summing junction (point "A") can be written as:

- $$I1 = I2 \qquad \text{eq. (4-4)}$$

We know from the properties of the ideal op amp that any voltage applied to the noninverting input (V_{in}) also appears at the inverting input. Therefore:

- $$V1 = V_{in} \qquad \text{eq. (4-5)}$$

From Ohm's law we know the value of current I1 is:

- $$I1 = \frac{V1}{R1} \qquad \text{eq. (4-6)}$$

— or, because $V1 = V_{in}$,

- $$I1 = \frac{V_{in}}{R1} \qquad \text{eq. (4-7)}$$

Similarly, current I2 is equal to the voltage drop across resistor R2, divided by the resistance of R1. The voltage drop across resistor R1 is the difference between the output voltage (V_o) and the voltage found at the inverting input (which is equal to V_{in}). Therefore:

- $$\frac{V_{in}}{R1} = \frac{V_o - V_{in}}{R2} \qquad \text{eq. (4-8)}$$

We home in on the transfer equation of the noninverting follower by substituting eqs. (4-7) and (4-8) into eq. (4-4). We must now solve eq. (4-9) for output voltage V_o:

- $$\frac{R2V_{vin}}{R1} = V_o - V_{in}$$ eq. (4-9)

- $$\frac{R2V_{in}}{R1} + V_{in} = V_o$$ eq. (4-10)

Factoring out V_{in}:

- $$V_{in}\left(\frac{R2}{R1} + 1\right) = V_o$$ eq. (4-11)

— or, reversing the order to the conventional style:

- $$V_o = V_{in}\left(\frac{R2}{R1} + 1\right)$$ eq. (4-12)

Equation (4-12) is the transfer equation for the noninverting follower. The voltage gain portion of eq. (4-12) should be either memorized or written down where you can easily find it.

Advantages of Noninverting Followers

The noninverting follower has several advantages. In our discussion of the unity gain configuration we mentioned that buffering and impedance transformation were advantages. Also, in the gain configuration we are able to provide voltage gain with no phase reversal — an advantage in some cases.

The input impedance of the noninverting followers shown thus far is very high, being essentially the input impedance of the op amp. In the ideal version, this impedance is infinite, while in practical devices it may range from 500K to more than 10^{12} ohms. Thus, the noninverting follower is useful for amplifying signals from any high impedance source, regardless of whether or not impedance transformation is a circuit requirement.

When the gain required is known (as it usually is) we select a trial value for R1, and then solve eq. (4-12) to find R2. This new version of the equation is:

- $$R2 = R1(A_v - 1)$$ eq. (4-13)

Using eq. (4-13) we can do simple design jobs.

Example

Design a gain-of-100 noninverting amplifier:

1. Select a trial value of resistance for R1 (typically 100 to 5000 ohms):

   R1 = 1200 ohms.

2. Use eq. (4-16) to determine the value of R2:

   $R2 = R1(A_v - 1)$
   $R2 = (1200)(100 - 1)$
   $R2 = (1200)(99)$
   $R2 = 118{,}800$

3. Determine whether or not the result obtained above is acceptable to you (if it is not, work the problem again). Now is the time for a little philosophizing. What does "acceptable" mean? If the value of R2 is exactly equal to a standard, easily obtained resistor, then all is well. However, as in the case above, the value 118,800 ohms is not a standard value. What we have to determine, therefore, is whether or not the nearest standard values result in an acceptable gain error (which is determined from the application). Both 118K and 120K are standard values, with 120K being somewhat easier to obtain. Both of these standard values are less than one percent from the calculated value, so this result is acceptable if the gain error with either or both standard values is within reasonable limits for the application. Figure 4-4 shows the resultant amplifier.

AC-Coupled Noninverting Amplifiers

The noninverting amplifiers discussed thus far have all been DC-coupled. In other words, they will respond to signals from DC up to the frequency limit of the amplifier selected. Sometimes, however, we do not want the amplifier to respond to DC or near-DC slowly varying signals. For these applications we

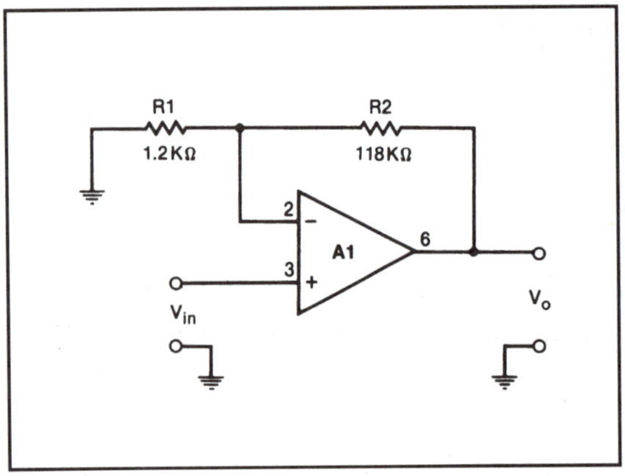

Figure 4-4. A gain-of-100 noninverting follower amplifier.

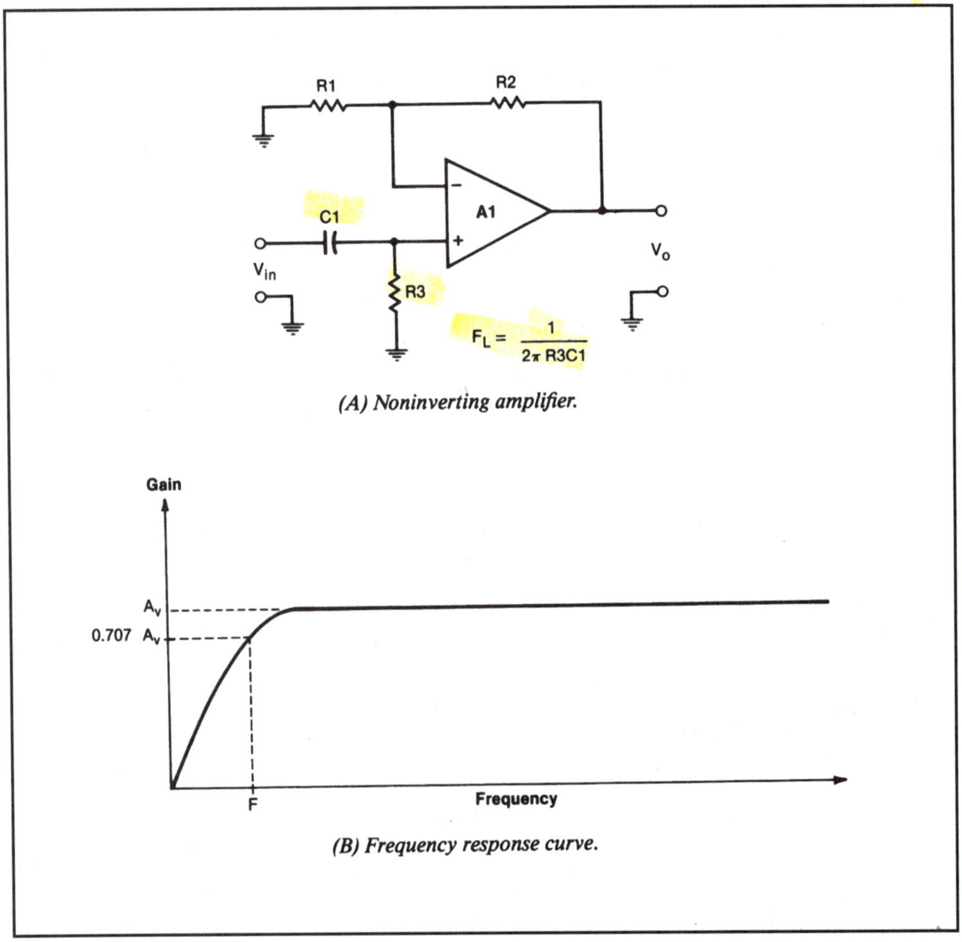

Figure 4-5. AC-coupled amplifier: A) circuit; B) frequency response.

select an AC-coupled noninverting follower circuit. In this section we will examine several AC-coupled noninverting amplifiers.

Figure 4-5A shows the simple capacitor-input AC-coupled amplifier circuit. It is essentially the same as the previous circuits, except for the input coupling network, C1/R3. The capacitor in Figure 4-5A serves to block DC and very low frequency AC signals. If the op amp has zero input bias currents, then we can delete resistor R3. Very few practical op amps even approach that ideal. As a result, input bias currents will charge capacitor C1, creating a voltage offset that is seen by the op amp as a valid DC signal and amplified to form an output offset voltage. In some devices, the output saturates from the C1 charge shortly after turn-on. Resistor R3 keeps C1 discharged.

Resistor R3 also sets the input impedance of the amplifier. Previous circuits had a very high input impedance because that parameter was determined only by the (extremely high) op amp input impedance. In Figure 4-5A, however, the Z_{in} seen by the source is equal to R3.

Another effect of resistor R3 and capacitor C1 is to limit the low frequency response of the circuit. This phenomena can be either a pain in the neck or a blessing. The blessing is derived from the fact that R3C1 form a high-pass filter (see Figure 4-5B). The -3 dB frequency, F, is found from:

- $$F = \frac{1{,}000{,}000}{2\pi R1C1} \qquad \textit{eq. (4-14)}$$

— where:

F is the -3 dB frequency in hertz (Hz).
R1 is the resistance in ohms (Ω).
C1 is the capacitance in microfarads (µF).

Example

Find the -3 dB frequency in hertz when C1 is 0.1 uF and R1 is 10 megohms.

- $$F = \frac{1{,}000{,}000}{(2)(3.14)(10{,}000{,}000\Omega)(0.1\mu F)} \qquad \textit{eq. (4-15)}$$

- $$F = 0.159 Hertz \qquad \textit{eq. (4-16)}$$

The form of eq. (4-14) is backward from what we normally need in practical design. In most cases, we know the required frequency response limit from the application. We also know from the application what the minimum value of R3 should be (an implication from source impedance), and often set it as high as possible as a practical matter (10 megohms is popular). Thus, we want to solve eq. (4-14) for C1 — as shown below:

- $$C = \frac{1{,}000{,}000}{2\pi R1F} \qquad eq.\ (4\text{-}17)$$

The technique of Figure 4-5 works well for dual-polarity DC power supply circuits. In single-polarity DC power supply circuits, however, the method falls down because of the large DC offset voltage present on the output. For these applications we use a circuit such as shown in Figure 4-6.

The circuit in Figure 4-6 is operated from a single V+ DC power supply (the V- terminal of the op-amp is grounded). In order to compensate for the V- supply being zeroed, the noninverting input is biased to a potential of:

- $$V1 = (V+)\left(\frac{R4}{R4+R5}\right) \qquad eq.\ (4\text{-}18)$$

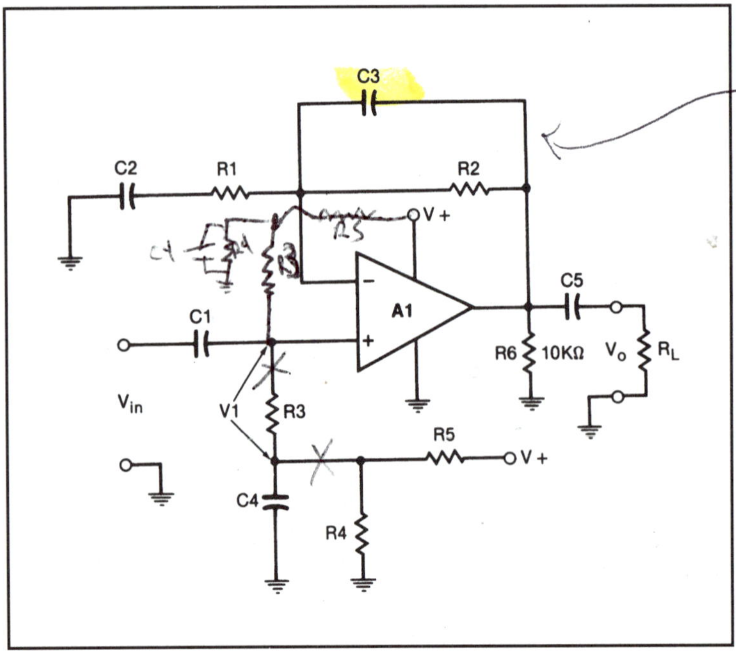

Figure 4-6. Single power supply operation of the AC amplifier.

If R4 = R5, then V1 will be (V+)/2. Because the noninverting input sinks no current, or very little current in the picoampere range, the voltage at both ends of R3 is the same (i.e., V1).

The circuit of Figure 4-6 does not pass DC and some low AC frequencies because of the capacitor coupling. Also, because capacitor C3 shunts feedback resistor R2, there is also a roll-off of the higher frequencies. The high frequency roll-off -3 dB point is found from:

$$F = \frac{1{,}000{,}000}{2\pi R2C3} \qquad \textit{eq. (4-19)}$$

— where:

F is the -3 dB frequency in hertz (Hz).
R2 is in ohms (Ω).
C3 is in microfarads (μF).

We can restate the “traditional” eq. (4-19) into a more useful form that takes into account the fact that we usually know the value of R2 (from setting the gain), and the nature of the application sets the minimum value of frequency F. We can rewrite eq. (4-19) in a form that yields the value of C3:

$$C3 = \frac{1{,}000{,}000}{2\pi R3F} \qquad \textit{eq. (4-20)}$$

The lower -3 dB frequency is set by any or all of several RC combinations within the circuit:

1. R1/C2
2. R3/C1
3. R3/C4
4. R_L/C5

Resistor R1 is part of the gain-setting feedback network. Capacitor C2 is used to keep the “cold” end of R1 above ground at DC, while keeping it grounded for AC signals.

Resistor R3 is the input resistor and serves the same purpose as the similar resistor in the previous circuit. At midband the input impedance is set by resistor R3, although at the extreme low end of the frequency range the reactance of C4 becomes a significant feature. In general, X_{C4} should be less than R3/10 at the lowest frequency of operation.

Capacitor C1 is in series with the input signal path and serves to block DC and certain very low frequency signals. The value of C1 should be:

- $$C1 = \frac{1{,}000{,}000}{2\pi FR3}$$ *eq. (4-21)*

— where:

C1 is in microfarads (µF).
F is in hertz (Hz).
R3 is in ohms (Ω).

Capacitor C5 is used to keep the DC output offset from affecting succeeding stages. The 10 K output load resistor (R6) keeps C5 from being charged by the DC offset voltage. The value of C5 should be greater than:

- $$C5 = \frac{1{,}000{,}000}{2\pi FR_L}$$ *eq. (4-22)*

— where:

C5 is in microfarads (µF).
F is the low-end -3 dB frequency in hertz (Hz).
RL is the load resistance in ohms (Ω).

Designer's Casebook Example

Design a gain-of-200 ±10 percent amplifier with -3 dB frequencies of 300 and 3300 hertz, and an input impedance of 2.2 megohms. Assume that the stage following will have a 10 K input impedance, so R_L for this design is 10,000 ohms. The V+ supply is 14.4 volts DC, and the noninverting input is biased to (V+)/2.

Procedure

1. Set the value of resistor R1 to 1K as a trial value.

2. Find the value of R2:

$$R2 = R1(A_v - 1)$$
$$R2 = (1K)(200 - 1)$$
$$R2 = (1K)(199)$$
$$R2 = 199K$$

The calculated value 199K is within 0.5 percent of standard value 200K. By specifying five percent tolerance resistors we ensure that the voltage gain is within tolerance. Set R1 = 1K, R2 = 200K.

3. Set the value of C2 such that X_{C2} is R1/10 or less. The minimum value of C2 is, therefore:

- $$C2_{min} = \frac{1{,}000{,}000}{2\pi F1R1} \qquad eq.\ (4\text{-}24)$$
- $$C2_{min} = \frac{1{,}000{,}000}{(2)(3.14)(300Hz)(1{,}000)ohms} \qquad eq.\ (4\text{-}25)$$
- $$C2_{min} = \frac{1{,}000{,}000}{1{,}884{,}000} = 0.54\mu F \qquad eq.\ (4\text{-}26)$$

The next higher standard value easily obtained by most hobbyists is 0.68 uF, with 1 uF being the next higher yet.

4. Calculate C3:

- $$C3 = \frac{1{,}000{,}000}{2\pi F2R2} \qquad eq.\ (4\text{-}27)$$
- $$C3 = \frac{1{,}000{,}000}{(2)(3.14)(3{,}300Hz)(200{,}000\ ohms)} \qquad eq.\ (4\text{-}28)$$
- $$C3 = \frac{1{,}000{,}000}{4{,}200{,}000{,}000} \qquad eq.\ (4\text{-}29)$$
- $$C3 = 0.00024\mu F = 240pF \qquad eq.\ (4\text{-}30)$$

The value 240 pF is a standard value, but is not usually stocked by local distributors. Use 220 pF if 240 pF is not easily available. Alternatively, construct a 240 pF capacitor from two smaller value capacitors (e.g., a pair of 120 pF in parallel).

5. Select R3 = 2.2 megohms to meet input impedance specification.

6. Select value for C4. The reactance of C4 should be R3/10 or less at frequency F1 (i.e., 300 Hz).

- $$C4 = \frac{1{,}000{,}000}{2\pi F1 X_{C4}}$$ eq. (4-31)
- $$C4 = \frac{1{,}000{,}000}{(2)(3.14)(300Hz)(220{,}000 ohms)}$$ eq. (4-32)
- $$C4 = \frac{1{,}000{,}000}{420{,}000{,}000} = 0.0025\mu F$$ eq. (4-33)

The value of C4 should be 0.0025 uF or more.

7. Set the value of C1:

- $$C1 = \frac{1{,}000{,}000}{2\pi R3 F1}$$ eq. (4-34)
- $$C1 = \frac{1{,}000{,}000}{(2)(3.14)(2{,}200{,}000\ ohms)(300Hz)}$$ eq. (4-35)
- $$C1 = 0.00024\mu F = 240pF$$ eq. (4-36)

8. Calculate the value of capacitor C5:

- $$C5 = \frac{1{,}000{,}000}{2\pi F1 R_L}$$ eq. (4-37)
- $$C5 = \frac{1{,}000{,}000}{(2)(3.14)(300Hz)(10{,}000\ ohms)}$$ eq. (4-38)
- $$C5 = \frac{1{,}000{,}000}{18{,}840{,}000} = 0.053\mu F$$ eq. (4-39)

9. Set values of resistors R4 and R5 if the bias voltage on the noninverting input is (V+)/2. In most cases, we will set these resistors equal to each other (for that bias), at a value between 1K and 10K. Use 2.2K = R4 = R5.

Summary

The following values apply for this amplifier:

R1 = 1000 ohms
R2 = 200,000 ohms
R3 = 2,200,000 ohms (2.2 megohms)
R4 = 2200 ohms
R5 = 2200 ohms
R6 = 10,000 ohms
C1 = 240 pF
C2 = 0.68 or 1.0 μF
C3 = 240 pF
C4 = 0.0025 μF
C5 = 1 μF

The values above will yield a gain-of-200 ±10 percent amplifier with a 2.2 megohm input impedance and a -3 dB bandwidth of 300 to 3,300 hertz.

Transformer-Coupled Noninverting Amplifier

Figure 4-7 shows the circuit for a transformer coupled noninverting follower. This type of circuit is often used in audio and broadcasting applications. For example, in those applications we might find audio signals passing over a 600-ohm balanced line. These circuits will be considered in greater detail later.

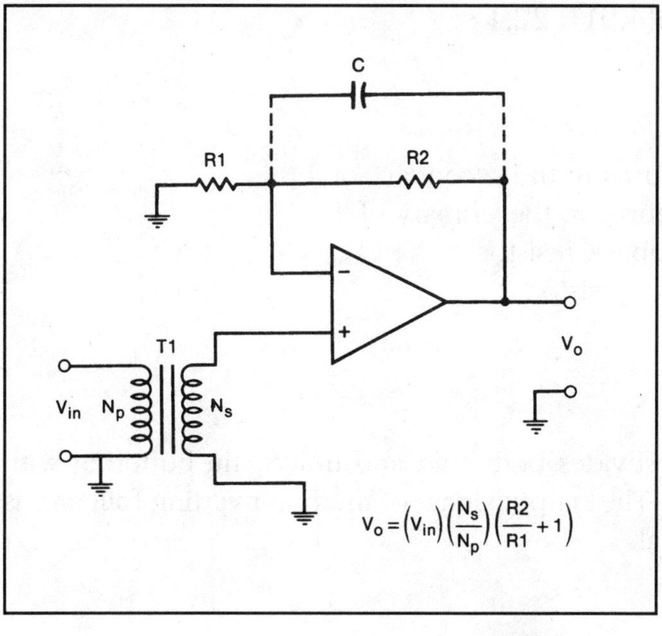

Figure 4-7. Transformer coupled AC amplifier.

The only point we will make here is that this circuit is an AC-only amplifier, with upper and lower -3 dB points determined mostly by (a) the frequency response of the transformer (T1), the limitations of the operational amplifier, and any capacitances shunting feedback resistor R2.

The gain of the amplifier in Figure 4-7 is given in eq. (4-40) in the example below.

Example

Find the gain of an amplifier such as Figure 4-7 if T1 has a turns ratio (Np/Ns) of 1:4.5 (which means that Ns/Np = 4.5:1), R1 = 1000 ohms and R2 = 3,900 ohms.

- $$A_v = (V_{in})\left(\frac{N_s}{N_p}\right)\left(\frac{R2}{R1} + 1\right) \qquad \textit{eq. (4-40)}$$

$$A_v = \left(\frac{4.5}{1}\right)\left(\frac{3{,}900}{1{,}000} + 1\right)$$

$$A_v = (4.5)(3.9) + 1$$

$$A_v = (4.5)(4.9) = 22.1$$

— where:

Av is the voltage gain.
Ns is the number of turns in the secondary of T1.
Np is the number of turns in the primary of T1.
R2 is the op amp feedback resistor.
R1 is the op amp input resistor.

Conclusion

The noninverting follower provides both gain and unity gain buffering with very high input impedances. The output signal of the noninverting follower is in phase with the input signal.

In the next chapter we'll take a look at the real world op amp. Up to this point, we've assumed that the op amp device met the ideal properties discussed in Chapter 2. While that approach makes it easy to derive transfer equations and explain how the thing works, the real world is different — some of those real op amps don't actually meet those ideals, believe it or not.

Chapter 5

Ideal vs. Practical Operational Amplifiers

The operational amplifiers that we considered in previous chapters were ideal devices; that is, they were textbook amplifiers, "perfect" in every way. That little ploy makes our equations work out nicely, but is somewhat naive for the real world. When you go to the electronics parts store (or order by mail-order), the kind of operational amplifiers that you will find are somewhat short of the ideal type enshrined in textbooks. In this chapter you will find a dose of reality — and the medicine needed to correct the problems that the operational amplifiers exhibit.

There are two main problems that we will consider. First, there are DC offsets on the output that are caused by any of several defects. We will discuss these different forms of offset, and provide several means for eliminating them. Second, we will examine the problem of frequency response. There are actually two aspects to this problem. One is gain-bandwidth product, which affects the maximum frequency response of the circuit, and the other is excess frequency response. The latter problem can lead to oscillation and ringing in the output circuit.

DC Offset Problems

A DC output offset is simply a DC voltage that appears on the output terminal not in response to any input signal. In other words, it is an output voltage that exists when it shouldn't. There are several sources of output offset voltage in real-world operational amplifiers.

Input Offset Current. One cause of the output offset voltage is input offset currents. Figure 5-1 shows the typical input stage used in bipolar operational amplifiers. Transistors Q1 and Q2 form an NPN differential pair. Although one of the ideal properties of the operational amplifier is that the input neither sinks nor sources current, real op amp inputs are transistors — and transistors need to be biased. If this were an ideal world, then the two currents would be equal and would cancel out. However, real op amps use transistors that are mismatched ever so slightly. The differential input bias current forms an input offset current:

$$I_{off} = I1 - I2 \qquad \text{eq. (5-1)}$$

The offset current (I_{off}) can produce an output voltage offset equal to the product of the current and the gain of the operational amplifier. While a low-gain op amp circuit may not exhibit a large output offset from the input offset current, high gain circuits almost invariably suffer such problems.

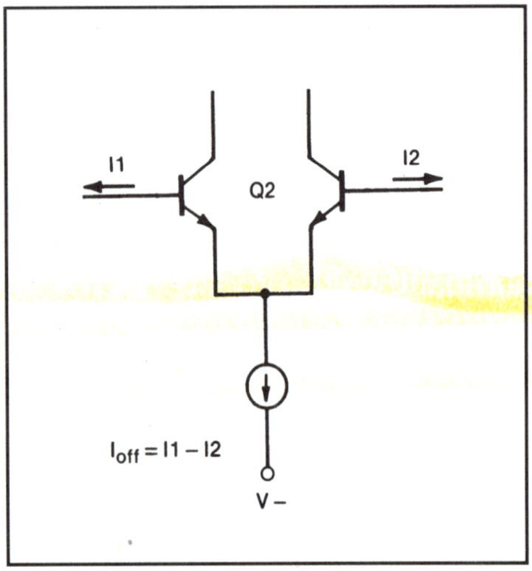

Figure 5-1. Bipolar transistor differential amplifier.

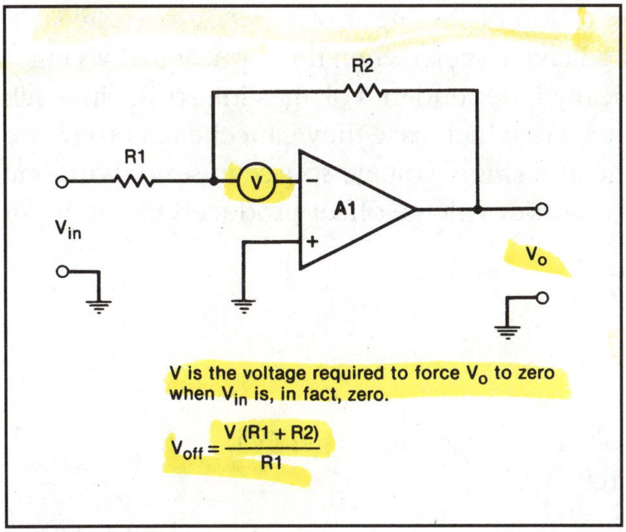

Figure 5-2. Input offset voltage model.

The severity of the problem depends in part upon the design of the operational amplifier. While certain old-fashioned types (e.g., the popular 741) have rather significant offset currents, certain newer types have reduced the current to a point where it is all but negligible. Especially low-input bias currents are found in those operational amplifiers that use either MOSFET transistors or JFET transistors for input transistors Q1 and Q2 (Figure 5-1). These devices are called BiMOS and BiFET operational amplifiers, respectively.

One problem with input offset currents is seen when a feedback network is used to set the gain of the operational amplifier (including both inverting and noninverting followers). The current flows in both input resistor (R1) and feedback resistor (R2), and produces an input offset voltage equal to the product of the current and the parallel combination of R1 and R2 (see Figure 5-2). By Ohm's law, this voltage is equal to:

$$V = \frac{I_{B1} R1 R2}{R1 + R2} \qquad eq.\ (5\text{-}2)$$

Below you will learn to cancel this voltage in practical circuits by using a compensation resistor in series with the other (i.e., noninverting) input.

Input Offset Voltage. The ideal operational amplifier has no voltage sources in series with either input except those external sources that supply signal. Real operational amplifiers, however, have an internal voltage source in series with one or both inputs. The definition of the input offset voltage is the voltage required to force the output voltage to zero when the input signal voltages are zero. Although there are typically independent voltages in series with each input, unless they are exactly equal (in which case they cancel each other) we can model the input offset voltage as a single voltage source in series with one of the two differential inputs. The output voltage offset produced by the input offset voltage is given by:

$$V_{out} = \frac{V(R1+R2)}{R1} \qquad \text{eq. (5-3)}$$

— where:

V_{out} is the output offset voltage at V_o in Figure 5-2 (V).
V is the input offset voltage (V).
R1 is the input resistance (Ω).
R2 is the feedback resistance (Ω).

Corrections For DC Offset Problems

Both input offset current and input offset voltage produce an output offset voltage that can adversely affect the operation of the amplifier. There are several ploys that we can use to eliminate the output offset voltage.

Figure 5-3 shows an inverting follower circuit that uses a compensation resistor (R3) to eliminate the output offset voltage that is caused by the input offset

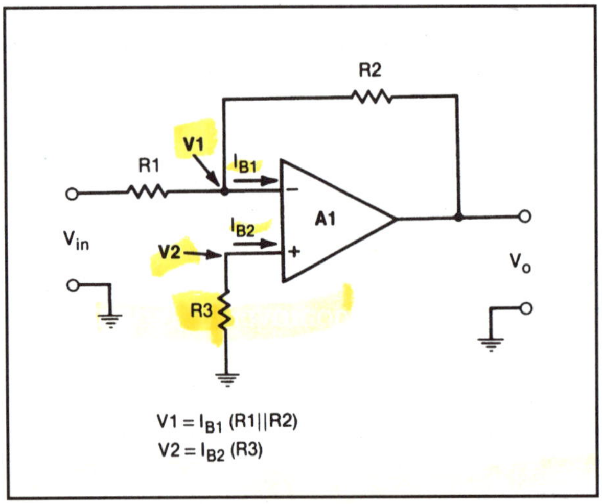

Figure 5-3. Use of compensation resistor to null offset voltage.

current. The input bias current flowing from the inverting input (IB1) will create a voltage drop (V1) across R1 and R2. This voltage was defined in eq. (5-2). Voltage V1 is seen by the op amp as a valid DC signal voltage. If the value of R3 is equal to the parallel combination of R1 and R2, then the voltage drop across it (V2) will be the same as the voltage applied to the inverting input. In other words, V1 = V2. This situation occurs because the two bias currents are almost equal in most operational amplifiers. The value of R3 is the parallel combination of R1 and R2, or:

$$R3 = \frac{R1R2}{R1+R2} \qquad \text{eq. (5-4)}$$

Figure 5-4 shows the use of built-in offset null terminals that are found on some operational amplifiers (typically pins 1 & 5 or 1 & 8). This circuit can eliminate output offset voltages from any source. The offset null potentiometer is connected between the offset null terminals, while the potentiometer wiper is connected to the negative DC power supply (V-). This method is preferred where the offset voltage is within the normal range of the offset null terminals.

The variant circuit shown in Figure 5-5 is used on the LM-101, LM-201 and LM-301 operational amplifiers. In this circuit the potentiometer is connected across the offset null terminals (as before), but the potentiometer wiper is connected to ground through a 5 megohm resistance (R4).

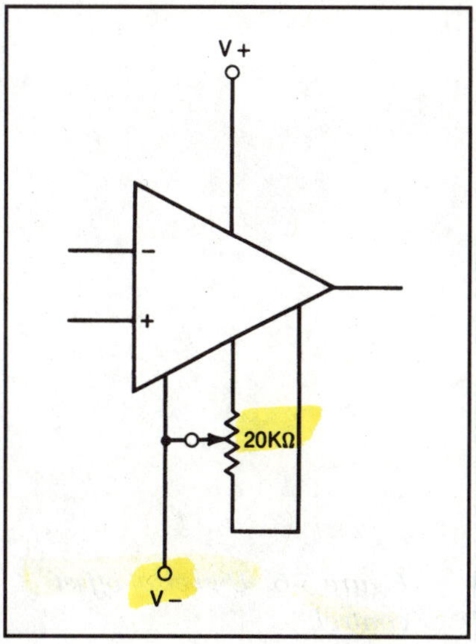

Figure 5-4. Offset potentiometer using op-amp null terminals.

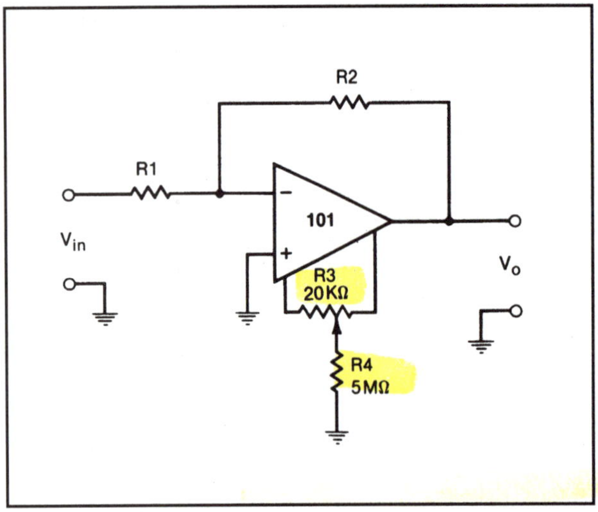

Figure 5-5. Alternate circuit using null terminals.

The potentiometer selected for any of these offset null circuits should be a ten- to twenty-turn trimmer potentiometer. This type of part allows maximum adjustability. The usual offset potentiometer is screwdriver operated, but if we use a shaft operated type, then we can use the same control as a position control for oscilloscopes, strip chart recorders and other purposes where an offset is intentionally selected.

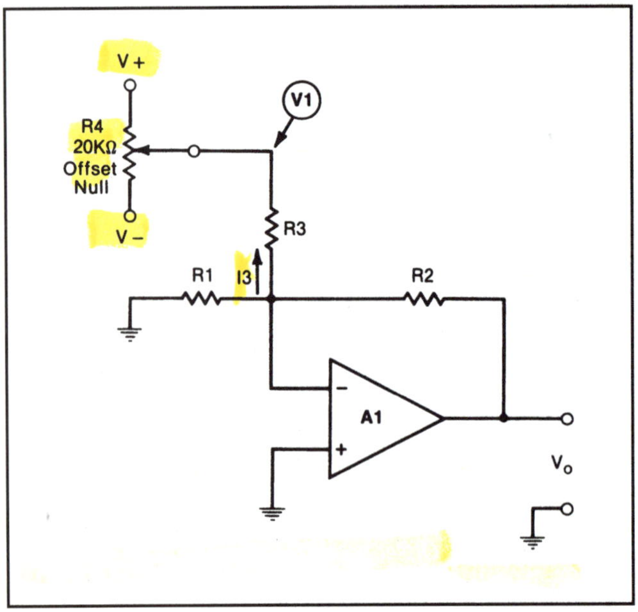

Figure 5-6. Universal offset control.

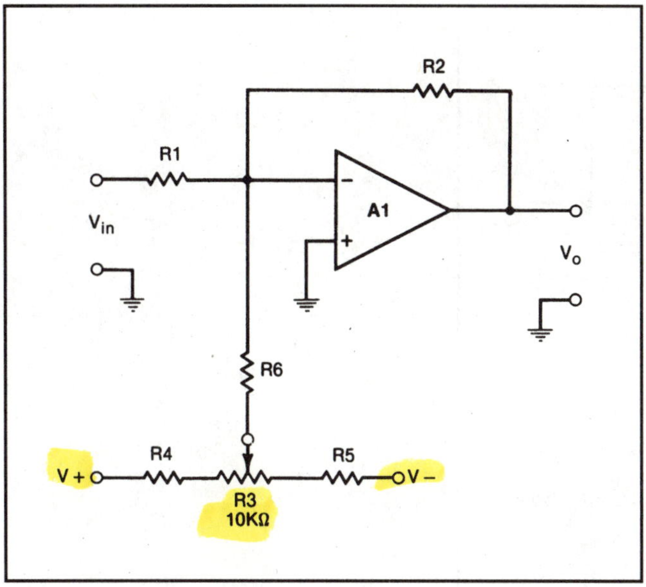

Figure 5-7. High resolution offset control.

Figures 5-6 through 5-9 show offset null circuits that can be used with any operational amplifiers whether or not they possess offset null terminals. The method shown in Figure 5-6 is, perhaps, the most commonly seen. In this case we cancel the output offset voltage by inserting a counter-current (I3) into the summing junction. When V1 is the voltage at the wiper of the offset null potentiometer the output voltage due to the adjustment of R4 is:

$$V_o = V1\left(\frac{-R2}{R3}\right) \qquad \text{eq. (5-5)}$$

The idea is to adjust potentiometer R4 to obtain an output voltage that exactly cancels the output offset. For example, if the output offset potential is +1.5 volts, then we need to select a value of V1 that makes eq. (5-5) equal to -1.5 volts. In many cases, we make R3 = R2, so the voltage at the wiper of potentiometer R4 (i.e., V1) should be equal to the output offset voltage that it cancels.

Figures 5-7 and 5-8 are variations on the same theme that permits better resolution adjustments of the offset voltage. These circuits are typically used in high-gain amplifier circuits with small signals. Those applications typically have very low-level signal inputs. The version in Figure 5-7 places two resistors in series with the arms of the potentiometer (R4 and R5). Typically the

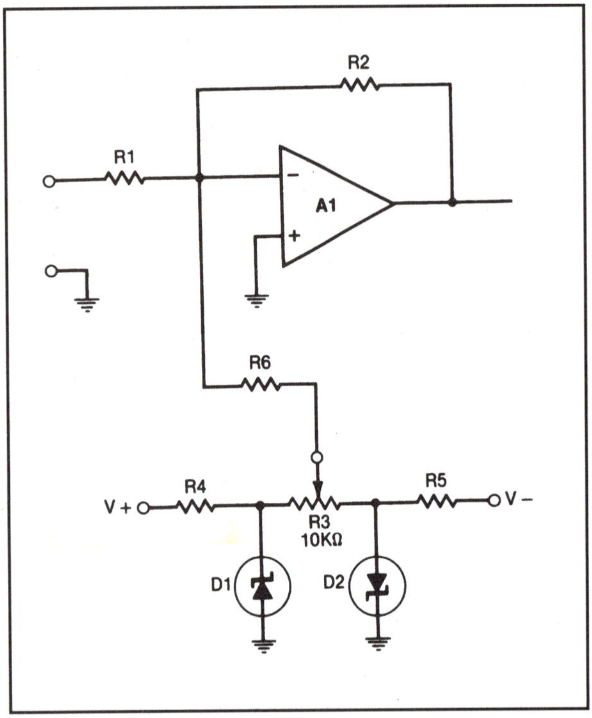

Figure 5-8. Alternate high resolution offset control.

value of R3 is smaller than the combined values of R4 and R5, and also R4 = R5. The version of Figure 5-8 uses a pair of zener diodes to place the ends of the potentiometer at lower potentials than the V- and V+ voltages that operate the amplifier. In most cases the zener potentials of D1 and D2 are equal, which gives a symmetrical range of offset voltages that can be canceled. There is no reason why different values can't be used if that is the situation of the specific circuit. One final offset null circuit is shown in Figure 5-9. In this case we apply the counter-voltage to the noninverting input rather than the inverting input. A small value resistor (R5) is connected between the noninverting input and ground. A resistor from the potentiometer wiper (R4) acts with R5 to form a resistor voltage divider. The voltage applied to the noninverting input (V2) is:

- $$V2 = \frac{V1R5}{R4+R5} \qquad eq.\ (5\text{-}6)$$

— or, given the value of R5:

- $$V2 = \frac{100}{R4+R5} \qquad eq.\ (5\text{-}7)$$

The output voltage produced by V2 acting on the operational amplifier's non-inverting input is:

$$V_o = V2\left(\frac{R2}{R1} + 1\right) \qquad \text{eq. (5-8)}$$

Because of the gain of the circuit we typically make V2 a very small value. We can use the equations above to select values for V1 and R4 required to cancel out the offset voltage.

The capacitor used between the inverting input and ground is optional in most cases. The purpose of C1 is to set the noninverting input at ground potential for AC signals, while keeping it at a non-zero value of V2 for DC. The value of C1 is such that it has a capacitive reactance of 10 ohms at the lowest frequency of operation. This capacitor can only be used in relatively low gain circuits, however, because one fault of op amps is that capacitors to ground from inputs increases the noise output from the circuit. In low gain circuits, however, the benefits may outweigh the noise problems.

Frequency Compensation Problems

The ideal amplifier of any type has an infinite bandwidth, so will amplify all signals accurately and cleanly. Real amplifiers, however, must contend with

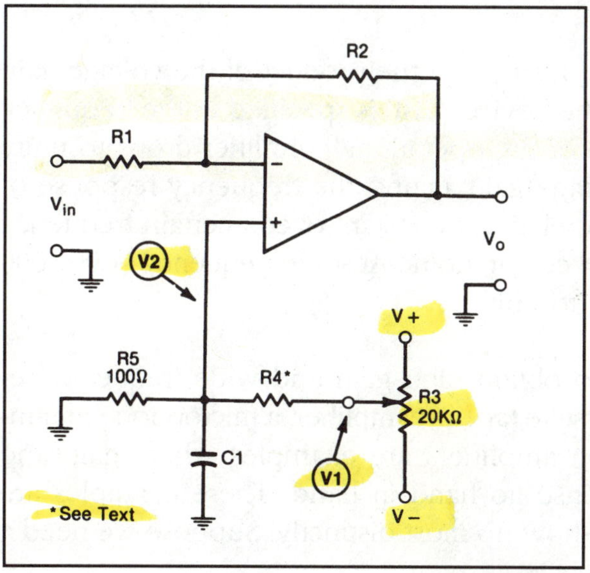

Figure 5-9. Offset control.

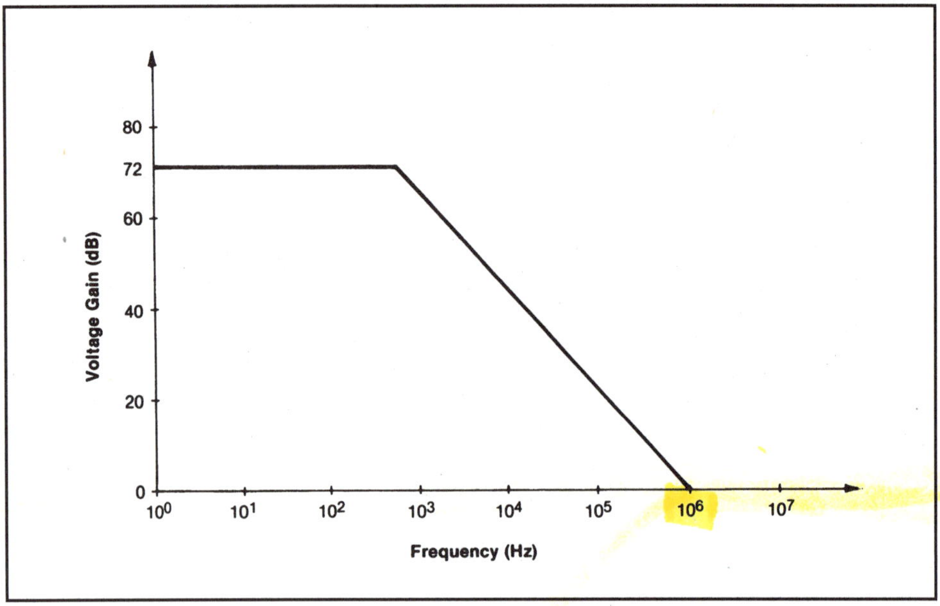

Figure 5-10. Gain-vs-frequency response.

frequency response limitations; for operational amplifiers these limitations can be quite severe. There are two problems to consider with regard to frequency response: the bandwidth of the system and the stability of the system. The frequency response problem requires us to understand the gain bandwidth product of the amplifier.

The gain bandwidth product (GBP) is merely the product of the voltage gain and the maximum bandwidth of the device. GBP is specified as the frequency in MHz or kHz at which the gain of the operational amplifier drops to unity (i.e., 1). Figure 5-10 plots the voltage gain against the frequency response of an amplifier. The gain will remain relatively constant over a certain frequency range, but will begin to fall above a certain point. At some frequency (1,000,000 hertz in this case), the gain drops to unity.

We run into GBP when we try to obtain high gain and wide frequency response in the same amplifier. A cassette tape preamplifier, a microphone preamplifier and certain instrumentation amplifiers are examples where gain and wide bandwidth frequency response go hand in hand. These examples are also where the problems of GBP show up most distinctly. Suppose we need a voltage gain of 40 dB (i.e., a gain of 100) and a frequency response of 10 kHz (i.e., 10,000 Hz). The gain bandwidth product required of the operational amplifier is:

- $GBW = A_v F$ *eq. (5-9)*
- $GBW = (100)(10,000Hz)$ *eq. (5-10)*
- $1,000Hz = 1MHz$ *eq. (5-11)*

The stability issue pops up whenever operational amplifiers are used. Almost invariably, when breadboarding a new circuit with wide-band operational amplifiers, the first attempt will result in an unwanted oscillation. One solution is to use a frequency compensated operational amplifier such as the 741 (if the application will permit). These devices are unconditionally stable — a condition that occurs at the expense of gain bandwidth product. Such an operational amplifier will work as an AC amplifier only to frequencies of a few kilohertz. These problems can be created from any of several problems. Layout, for example, can create oscillation. If the stray capacitances and inductances create a resonant situation at a frequency lower than the unity gain frequency, then the circuit will oscillate. Similarly, if the DC power supply is not properly decoupled (see Chapter 2) then problems can occur. This latter problem especially occurs in multistage circuits where one stage can affect the other. Be careful of disc ceramic capacitors, however, for some of those components have very significant stray inductances and can thus form resonant circuits that do not bypass very well at one frequency. If that resonant frequency is below the GBP frequency, then oscillation may occur.

Grounding problems in the circuit can also create a problem. One cause of oscillations in these cases is an impedance in the ground circuit that is common to both input and output circuits. Because output signal flowing in that impedance is also in series with the input circuit, feedback automatically exists.

If any combination of feedback and phase shift occurs between output and input, at a frequency that is less than the unity gain GBP frequency, then oscillation can occur. Diagnosing the source of oscillations in op amp circuits

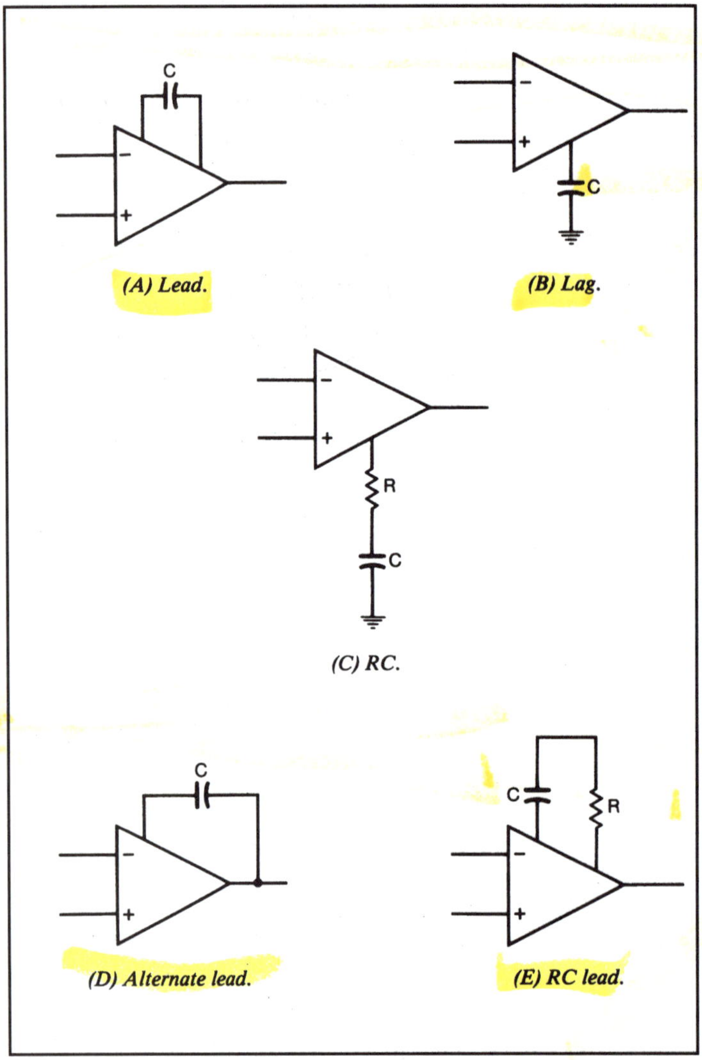

Figure 5-11. Frequency compensation circuits: A) lead; B) lag; C) R-C; D) alternate lead; E) R-C lead.

is sometimes a bit difficult. Possible causes include external interference, DC power supply ripple reduction, decoupling between cascade stages through the DC power lines, layout, ground loops, and so forth.

There is no real good rule for figuring out where the problem is located, although experienced troubleshooters often develop a "sixth sense" about what to look at first. There is a rule of thumb, however. Measure the oscillation frequency (F_o) and compare it with the gain-bandwidth product (F_t) found in the data sheet. If F_o is close to F_t, then it's probable that the problem is in the feedback network. A diagnostic aid is to temporarily increase the closed-loop gain of the amplifier by a factor between two and ten (2-10). Observe the effect on stability. If the oscillation ceases, or if oscillation frequency F_o drops significantly, then the problem is probably in the feedback network. If neither of these events occurs, then the problem is elsewhere.

There are several ploys that are used to reduce the gain at frequencies less than GBP to prevent oscillation. Some of these methods are shown in Figure 5-11. In Figure 5-11A we see the use of a capacitor across the frequency compensation terminals found on some op amps. The manufacturer typically specifies a capacitance in their data sheet for the device. Values in the 30 to 100 pF are common. Standard practice is to make this capacitance the value for the noninverting unity gain follower configuration. For all other configurations with gain, the value of the capacitance required is the noninverting unity gain value multiplied by the feedback factor, Ω:

- $$C_{req} = C\beta \qquad \text{eq. (5-12)}$$

— where:

C_{req} is the required capacitance (pF).
C is the specified capacitance for noninverting unity gain configuration (pF).
β is the feedback factor

The factor β is merely the transfer function of the feedback network. In the usual circuit with a resistor feedback network, the value of β is:

- $$\beta = \frac{R1}{R1+R2} \qquad \text{eq. (5-13)}$$

— where:

β is the feedback factor.
R1 is the input resistor (Ω).
R2 is the feedback resistor (Ω)

For example, let's suppose that an inverting amplifier with a gain of 10 has a rated compensation capacitance of 30 pF for the unity gain configuration. Assume the feedback resistor (R2) is 100K, and the input resistor (R1) is 10K.

The value of β is:

$$\beta = \frac{10K}{10K + 100K} \qquad \text{eq. (5-14)}$$

$$\beta = \frac{10}{110} = 0.091$$

The value of the capacitance required to stabilize this amplifier is:

$$C_{req} = C\beta \qquad \text{eq. (5-15)}$$

$$C_{req} = (30pF)(0.091) \qquad \text{eq. (5-16)}$$

$$C_{req} = 2.7pF \qquad \text{eq. (5-17)}$$

Figures 5-11B through 5-11E show variations on the theme. The values of the components are typically taken from charts or graphs in the specifications sheet for the particular operational amplifier being used.

Input Capacitance Compensation. Figure 5-12 shows the method for compensating for stray input capacitance. The circuit shown here is an inverting follower with a capacitance shunting the feedback resistor. From DC to a given frequency, F1, the gain of the amplifier is given by the equation:

$$A_v = \frac{-R2}{R1} \qquad \text{eq. (5-18)}$$

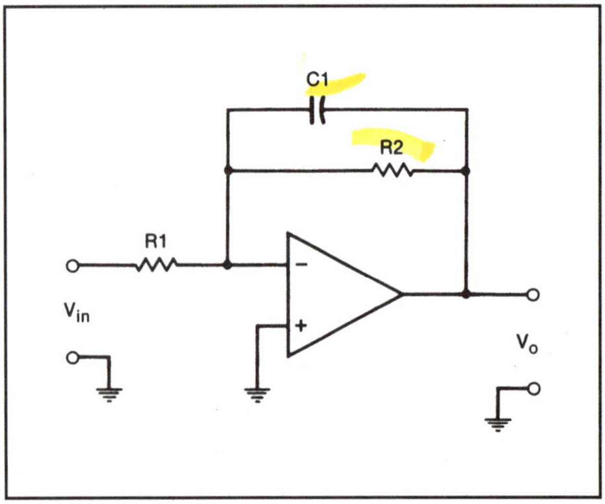

Figure 5-12. Feedback capacitor controls frequency response on upper end of spectrum.

At frequency F1, however, the gain begins to drop off at a rate of -6 dB per octave. The breakpoint frequency (F1) at which this happens is:

$$F1 = \frac{1{,}000{,}000}{2\pi R2C1} \quad \text{eq. (5-19)}$$

— where:

F1 is the frequency in Hertz (Hz).
R2 is in ohms (Ω).
C1 is in microfarads (μF).

The input resistance and the input capacitance (which consists mostly of stray capacitances) forms a phase shift at certain frequencies defined by 1/2πR1C, where C is the stray input capacitance. If that frequency is below the unity gain frequency, then it is possible for the phase-shift that occurs at this frequency to add to the operational amplifier's ordinary phase shift (180 degrees ± error for inverting followers) to form a 360-degree phase-shift, which is the criteria for oscillation. In order to compensate for this stray capacitance select a value of C1 (Figure 5-12) that has a capacitive reactance of approximately 1/10th the resistance of R2.

Figure 5-13 shows the method for compensating for high output load capacitances. This situation is found when the operational amplifier drives a coaxial cable or twisted pair transmission line (and other places). Capacitor C1 is selected to have a capacitive reactance of 1/10th R2, while R3 is selected to have a value of several hundred ohms.

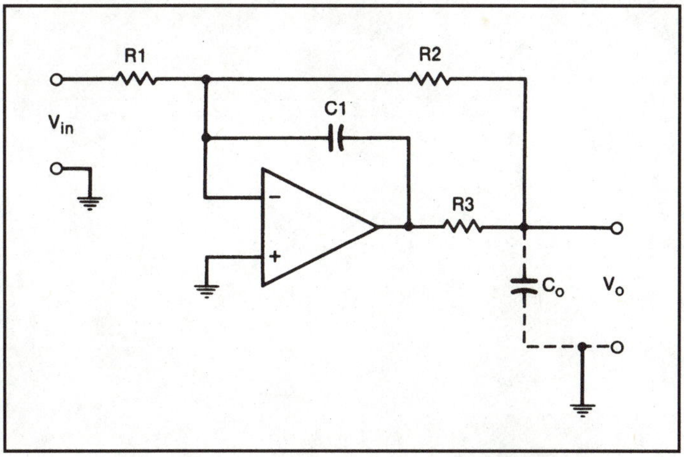

Figure 5-13. Alternate feedback capacitor circuit.

Chapter 6

DC Differential Operational Amplifier Circuits

Most operational amplifiers have differential inputs, even though they are not required to by the most fundamental op amp definition. That is, there are two inputs on the device that provide the same amount of gain but have opposite polarity sense. The inverting input (-IN) of the operational amplifier provides an output that is 180 degrees out of phase with the input signal. In other words, a positive-going input signal will produce a negative-going output signal, and vice-versa. The noninverting input (+IN) produces an output signal that is in phase with the input signal. For this type of input, a positive-going input signal will produce a positive-going output signal.

Common Mode Signals

The use of differential inputs is shown in Figure 6-1. First let's consider the "common mode" signal, V3. A common mode signal is one which is applied to both inputs at the same time. Such a signal might be either a voltage such as V3, or, a case where voltages V1 and V2 are equal to each other (i.e., V1 = V2). The implication of the common mode signal is that, being applied equally to inverting and noninverting inputs, the output voltage is zero. Because the two

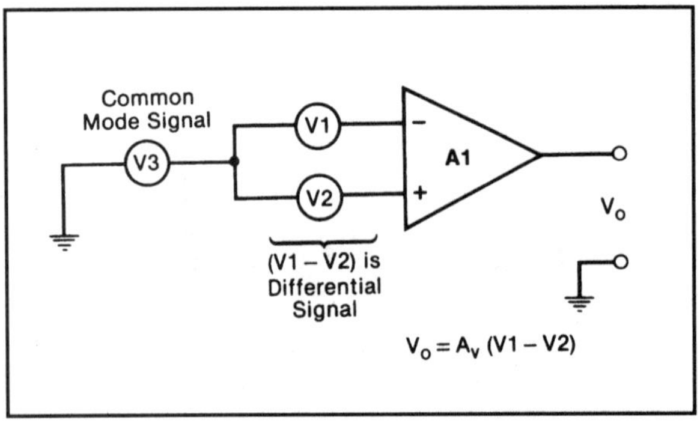

Figure 6-1. Signal inputs to a differential amplifier.

inputs have equal but opposite polarity gain for common mode signals, the net output signal is zero.

The operational amplifier with differential inputs will cancel common mode signals. An example is in the performance of the differential amplifier with respect to 60 Hz hum pick-up. Almost all input signal cables for practical amplifiers will pick up 60 Hz energy and convert it to a voltage that is seen as a genuine input signal. In a differential amplifier, however, the 60 Hz field will affect both inverting and noninverting inputs equally; so the 60 Hz artefact signal will disappear in the output. Neat trick, huh?

The practical operational amplifier will not have a perfect rejection of common mode signals. A specification called the Common Mode Rejection Ratio (CMRR) tells us something of the ability of any given op amp to reject such signals. The CMRR is usually specified in decibels (dB), and is defined as:

$$CMRR = \frac{A_v}{A_{cm}} \qquad \text{eq. (6-1)}$$

— where:

CMRR is the common mode rejection ratio.
A_v is the voltage gain to differential signals.
A_{cm} is the voltage gain to common mode signals.

In general, the higher the CMRR the better the operational amplifier. Typical low-cost devices have CMRR ratings of 60 dB or more.

Differential Signals

Signals V1 and V2 in Figure 6-1 are differential signals. The total differential signal seen by the operational amplifier is:

- $$V_d = V2 - V1$$ *eq. (6-2)*

The output signal from the differential operational amplifier is the product of the differential voltage gain and the difference between the two input signals (hence the term "differential" amplifier). Thus, the transfer equation for the operational amplifier is:

- $$V_o = A_v(V2 - V1)$$ *eq. (6-3)*

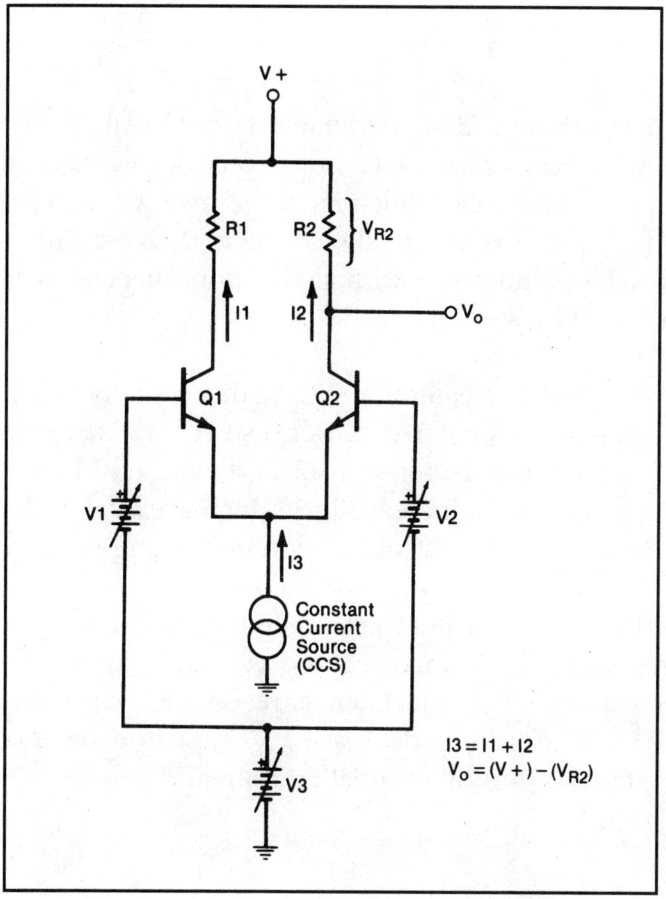

Figure 6-2. DC differential op-amp input circuit.

Differential Input Stage

Figure 6-2 shows a hypothetical input stage for a differential input operational amplifier. Transistors Q1 and Q2 are a matched differential pair that share a common DC power supply through separate collector load resistors. The collector of transistor Q2 serves as the output terminal for the differential amplifier. Thus, collector voltage V_o is the output voltage of the stage. The exact value of the output voltage is the difference between supply voltage (V+) and the voltage drop across resistor R2 (i.e., VR2). The emitter terminals of the two transistors are connected together, and are fed from a constant current source (CCS). For purposes of analysis, we can assume that the collector currents (I1 and I2) are equal to each other. While this convention might choke a transistor purist, it is nonetheless useful for our purposes. Because of Kirchoff's Current Law (KCL), we know that:

$$I3 = I1 + I2 \qquad \textit{eq. (6-4)}$$

Because current I3 is a constant current, a change in either I1 or I2 will change the other. For example, an increase in current I1 must force a decrease in current I2. Now let's consider how the stage works. First, the case where V1 = V2. In this case, both Q1 and Q2 are biased equally, so the collector currents (I1 and I2) are equal to each other. Under the normal situation, this case will result in VR2 being equal to V_o, and also equal to (V+)/2.

Next, let's consider the case where V1 is greater than V2. In this case, transistor Q1 is biased on harder than Q2, so I1 will increase. Because I1 + I2 is always equal to I3, an increase in I1 will force a decrease in I2. A decrease of I2 will also reduce VR2 and increase V_o. Thus, an increase in V1 increases V_o, so the base of transistor Q1 is the noninverting input of the differential amplifier.

Now let's consider the case where V2 is greater than V1. In this case, transistor Q2 is biased on harder than Q1 so I2 will increase and I1 will decrease. An increase in I2 forces a larger voltage drop (VR2) across resistor R2, so V_o will go down. Thus, an increase in V2 forces a decrease in V_o, so the base of transistor Q2 is the inverting input of the differential amplifier.

Voltage V3 is a common mode signal, so will affect transistors Q1 and Q2 equally. For this kind of signal the output voltage (V_o) will not change.

The base bias currents required to keep transistors Q1 and Q2 operating become the input bias currents that make the practical op amp non-ideal. In order to make the input impedance high we need to make these currents very low. Some manufacturers offer operational amplifiers with MOSFET transistors (called BiMOS op amps) or JFET transistors (BiFET op amps). These types of transistors inherently have lower input bias currents than bipolar NPN or PNP transistors, and in fact the bias currents approach mere leakage currents.

Practical Circuits

Figure 6-3 shows the circuit of the simple DC differential amplifier based on the operational amplifier device. The gain of the circuit is set by the ratio of two resistors:

- $$A_v = \frac{R3}{R1} \qquad \textit{eq. (6-5)}$$

— or,

- $$A_v = \frac{R4}{R2} \qquad \textit{eq. (6-6)}$$

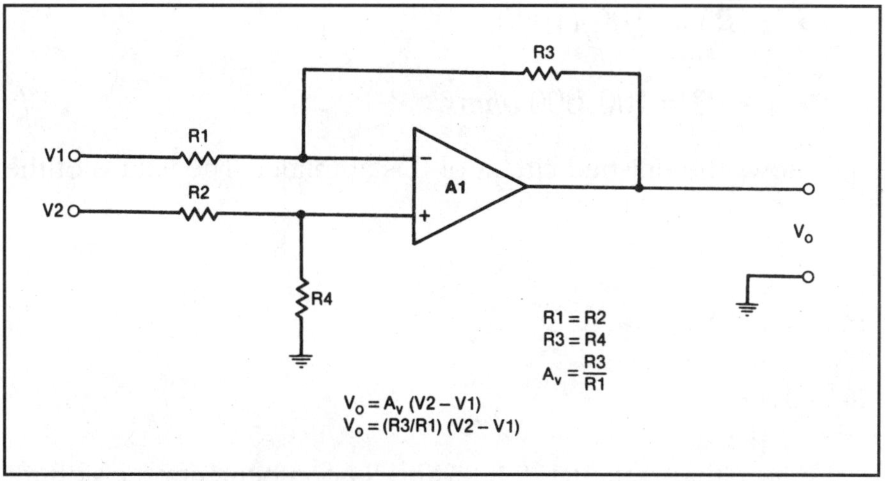

Figure 6-3. Simple DC differential amplifier.

Provided that R1 = R2 and R3 = R4.

The constraint is that R1 = R2, and R3 = R4. These balances must be maintained or the common mode rejection ratio (CMRR) will deteriorate rapidly. In most common applications, the CMRR can be maintained within reason by specifying five percent-tolerance resistors for R1 through R4. However, where superior CMRR is required, especially where the differential voltage gain is high, closer tolerance resistors (one percent or better) are required.

Designer's Casebook Example

Design a DC differential amplifier with a gain of 100. Assume that the source impedance of the preceding stage is about 100 ohms.

1. Because the source impedance is 100 ohms, we need to make the input resistors of the DC differential amplifier ten times larger (or more). Thus, the input resistors (R1 and R2) must be 1000 ohms: R1 = R2 = 1000 ohms.

2. Set the value of the two feedback resistors (keep in mind that R3 = R4). The value of these resistors is found from rewriting eq. (6-5):

- $R3 = A_v R1$ *eq. (6-7)*
- $R3 = (100)(1000)$ *eq. (6-8)*
- $R3 = 100,000\ ohms$ *eq. (6-9)*

Figure 6-4 shows the finished circuit of this amplifier. The values of the resistors are:

R1 = 1K
R2 = 1K
R3 = 100K
R4 = 100K

For best results, either make R1, R2, R3 and R4 one percent precision resistors, or hand-match five percent resistors on an ohmmeter.

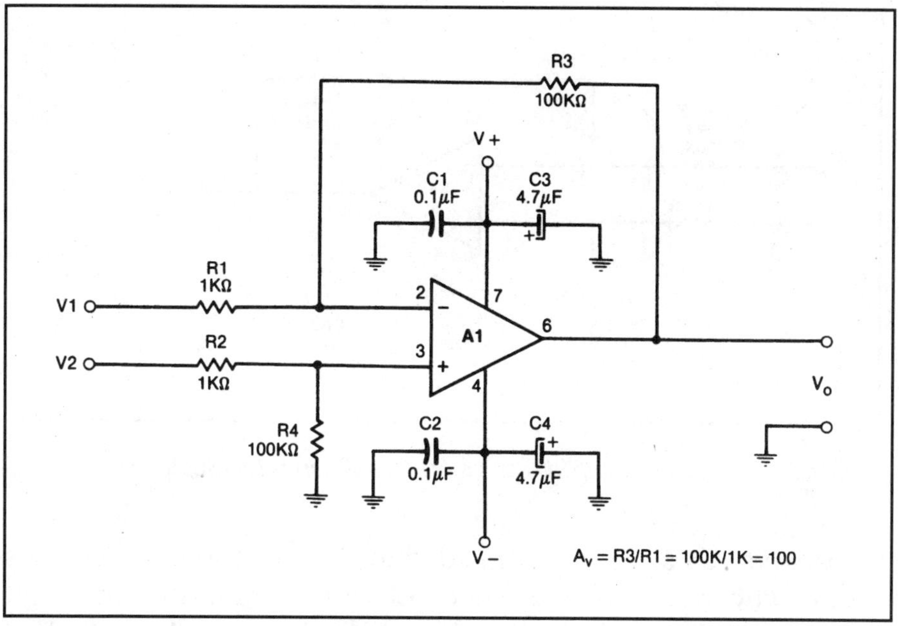

Figure 6-4. Gain-of-100 DC amplifier.

The pinouts shown in Figure 6-4 are for the "industry standard" 741-family of operational amplifiers. These same pinouts are found on many different op amp products.

The DC power supply voltages are usually either ±12 volts DC, or ±15 volts DC. Lower potentials can be accommodated, however, if a corresponding reduction in the output voltage swing is tolerable. A typical op amp will produce a maximum output voltage that is approximately 3 volts lower than the supply voltage. For example, when V+ is 12 volts DC the maximum positive output voltage permitted is (12 - 3), or 9 volts. We can use ±9 volts (and take advantage of 9 volt transistor radio batteries) for the DC power supplies, and that would limit the maximum signal output voltage to ±6 volts. Some BiFET and BiMOS op amps can swing to within 0.5 volts of the DC power supply rail.

The decoupling bypass capacitors shown in Figure 6-4 are used to keep the circuit stable, especially in those cases where the same DC power supplies are used for several stages. The low value 0.1 μF capacitors (C1 and C2) are used to decouple high frequency signals. These capacitors should be physically placed as close as possible to the body of the operational amplifier. The high

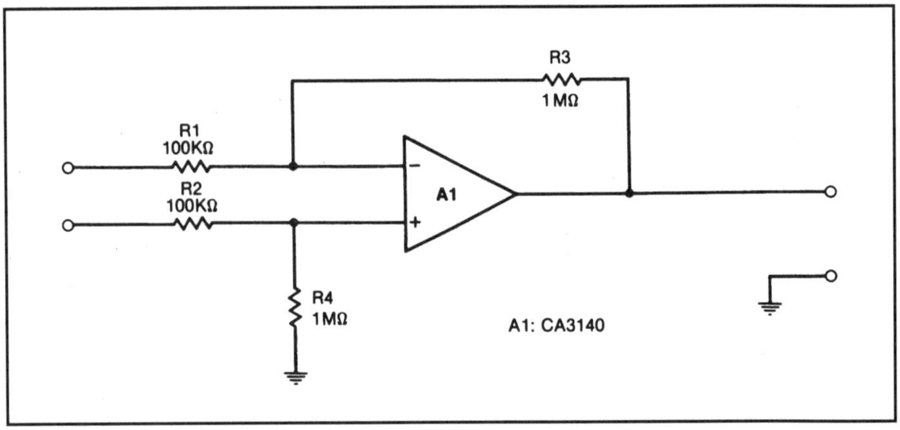

Figure 6-5. High input impedance gain-of-10 differential amplifier.

value capacitors (C3 and C4) are needed to decouple low frequency signals. The reason why we need two values of capacitors is that the high value capacitors needed for low frequency decoupling are typically electrolytics (tantalum or aluminum), which are ineffective at high frequencies. Thus, we must provide smaller value capacitors that (1) have a low enough capacitive reactance to do the job, and (2) are effective at high frequencies.

A different situation is shown in Figure 6-5. Most differential amplifiers have relatively low input impedances, which is a function of factors such as input bias currents and so forth. The amplifier in Figure 6-5 has a high input impedance by virtue of the high values of resistors R1 and R2. In order to attain this input impedance, however, we need to specify an operational amplifier for A1 that has a very low input bias current (i.e., a very high natural input impedance). The BiMOS device, which uses MOSFET input transistors, and the BiFET, which uses JFET input transistors, are good selections for DC differential amplifier circuits with high input impedances.

CMRR Adjustment

The common mode rejection ratio (CMRR) of the operational amplifier DC differential circuit is dependent upon two main factors. First, the natural CMRR of the operational amplifier used as the active device. Second, the balance of the resistors, R1 = R2 and R3 = R4. Unfortunately, the balance is typically

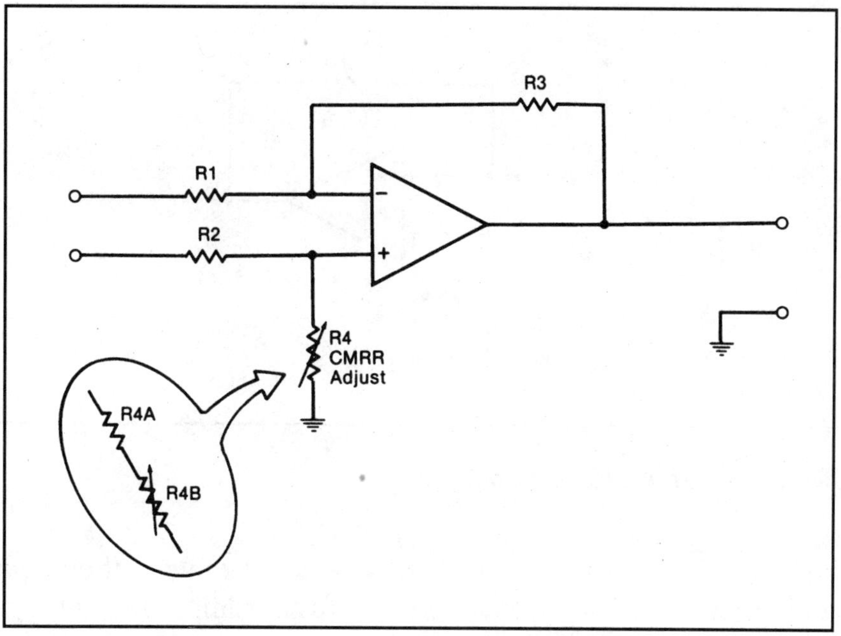

Figure 6-6. CMRR adjustment control added to the differential amplifier.

difficult to obtain with fixed resistors. We can use a circuit such as Figure 6-6. In this circuit, R1 through R3 are exactly the same as in previous circuits. The fourth resistor, however, is a potentiometer (see R4). The potentiometer will "adjust out" the CMRR errors caused by resistor and related mismatches.

A version of the circuit with greater resolution is shown in the inset to Figure 6-6. In this version the single potentiometer is replaced by a fixed resistor and a potentiometer in series, the sum resistance (R4A + R4B) is equal to about twenty percent more than the normal value of R3. Ordinarily, the maximum value of the potentiometer is ten to twenty percent of the overall resistance. The adjustment procedure for either version of Figure 6-6 is the same (see Figure 6-7):

1. Connect a zero-center DC voltmeter to the output terminal (M1 in Figure 6-7). Short together inputs A and B, and then connect them to either a moderate voltage source, or ground. Adjust potentiometer R4 (CMRR ADJ) for zero volts output.

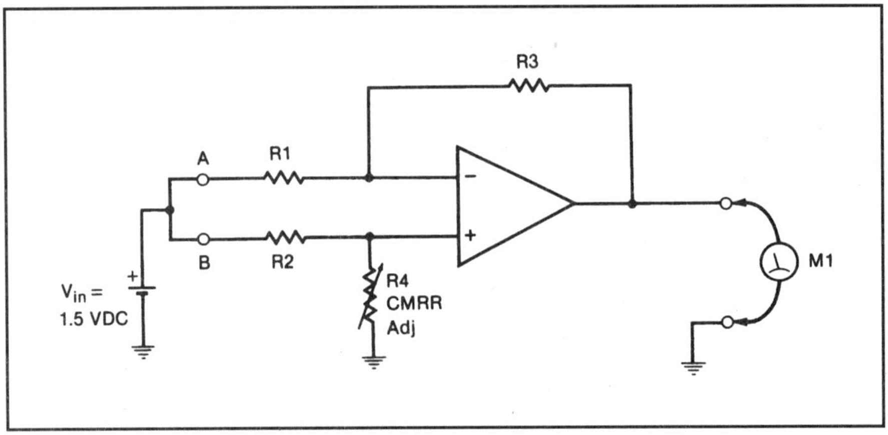

Figure 6-7. CMRR adjustment test circuit.

2. If the output indicator (meter M1) has several ranges, then switch to a lower range and repeat step above until no further improvement is possible.

Alternatively, connect the output to an audio voltmeter or oscilloscope, and connect the input to a 1-volt peak-to-peak AC signal that is within the frequency range of the particular amplifier. For audio amplifiers, a 400 to 1000 hertz, 1-volt signal is typically used.

Universal Diff-Amp Project

Now that we have the basic differential amplifier down pat, let's take a look at a simple printed circuit board project that can actually be used. Figure 6-8A shows the circuit for the amplifier, while Figure 6-8B shows the 1:1 printed circuit pattern. The printed circuit pattern can be used to make a printed circuit board using any of several "home brew" processes (see local electronic parts distributors or the mail-order houses for PCB kits). Alternatively, you can cut the pattern out and send it to an printed circuit maker.

The circuit of the differential amplifier in Figure 6-8A is the standard circuit discussed above. The component values are not specified, and can be derived using the procedures discussed above. The pinouts of the op amp are the industry standard, so any device can be used if it meets this specification.

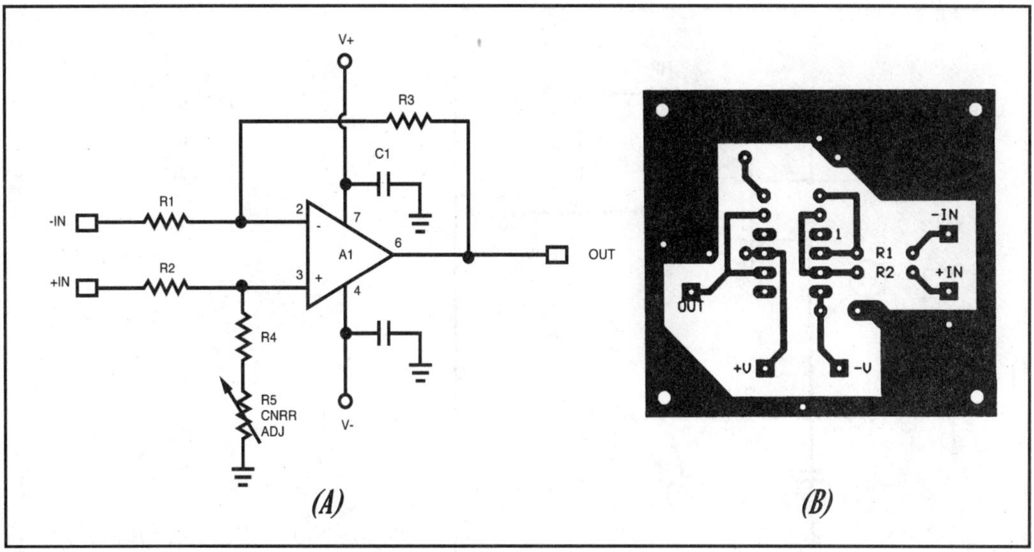

Figure 6-8. "Universal" DC differential amplifier project: A) circuit, B) printed circuit pattern.

Examples include the 741 for compensated op amps, and the CA-3140 in the mini-DIP package for wider bandwidth devices.

CMRR adjustment is provided by a potentiometer (R5) in series with R4. The value of R4 + R5 should be equal to R3 for best CMRR. It is common practice to make R4 about 80 percent of R3, and then select a potentiometer that has a range that will set the resistance of the combination (R4+R5) equal to about R3 ± 20 percent or so. To make the CMRR adjustment, follow the procedure outlined above.

Variable Gain Differential Amplifier

The standard "simple" DC differential amplifier circuit discussed in this chapter suffers from a major problem: it is hard to make the voltage gain variable. To make the gain variable, it would be necessary to use a dual potentiometer connected such that one resistor is in the feedback loop of the op amp, and the other goes from the +IN terminal to ground. Not too neat, especially when the CMRR depends on those two potentiometers tracking with respect to resis-

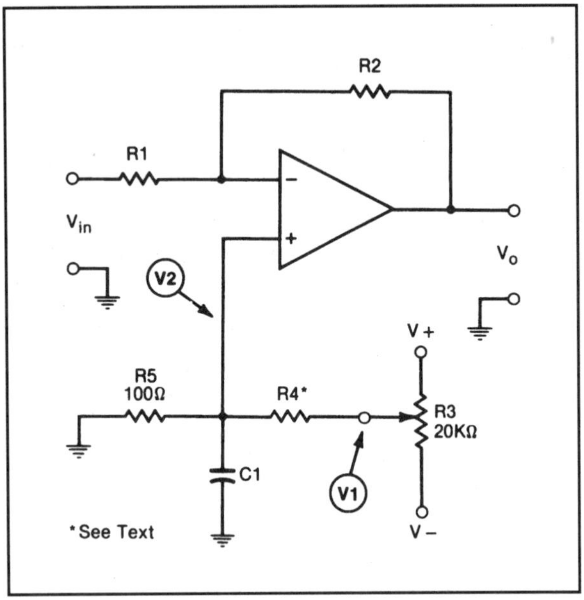

Figure 6-9. DC differential amplifier with single gain control.

tance value as the shaft it turned. However, there is another way; consider Figure 6-9.

The circuit of Figure 6-9 modifies the differential amplifier configuration to provide a gain control potentiometer. Note that both the feedback resistance (R6 and R7 from output to -IN) and the balancing CMR resistance (R3 and R4) are broken into two equal segments. The gain control potentiometer (R5) is connected between the nodes created by R6/R7 and R3/R4. The gain of this circuit is said to be:

$$A_v = \frac{2R3}{R1\left(\frac{1+R3}{R5}\right)} \qquad \text{eq. (6-10)}$$

Recommended values for this circuit: R1 and R2 should be 15 or 18K, while all other resistors should be 100K, including potentiometer R5.

Some Special Diff-Amps

All of the previous circuits have been based on the discrete operational amplifier equipped with differential inputs (-IN and +IN). In this section we take a look at a couple of products that have the resistors built in.

The Burr-Brown INA-105 device is shown in Figure 6-10. This device is a precision unity gain differential amplifier that provides a CMRR of at least 86 dB over a wide temperature range. It has a gain error of only 0.01%, and a linearity of 0.001%. It can be used as a gain-of-2, gain-of-1 or gain-of-0.5 amplifier depending on how the pins are cross-connected. It can also be used as a 4-to-20 mA current source transmitter (used in instrumentation applications), or as a voltage-controlled current source (which amounts to about the same thing). Its applications include buffering, average value generation, or a filter (again, depending on how the pins are connected).

In the Next Chapter

The simple DC differential amplifier circuit shown in this chapter is useful for low gain applications, and for those applications where a low to moderate input impedance is permissible (e.g., 300 to 200,000 ohms). Where a higher gain is required, then we must resort to a more complex circuit called the operational amplifier instrumentation amplifier, or "IA." In the next chapter we will examine the classical three-op amp IA circuit, as well as several integrated circuit instrumentation amplifiers (ICIA) that offer the advantages of the IA in a single small IC package. Some of those devices are now among the most commonly used in many instrumentation applications.

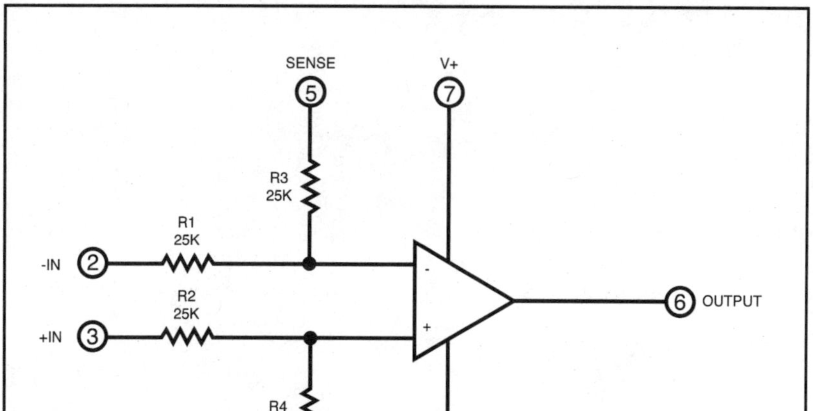

Figure 6-10. Burr-Brown INA-105 differential amplifier IC.

Chapter 7

Instrumentation Amplifiers

The simple DC differential amplifier discussed in Chapter 6 suffers from several important drawbacks. First, there is a limit to the input impedance (Z_{in} is approximately equal to the sum of the two input resistors). Because the input impedance must not be less than ten times the source impedance, you can reduce the input impedance only so far before running into problems.

Second, there is a practical limitation on the gain available from the simple single-device DC differential amplifier of Chapter 6. If high gain is attempted, then either the input bias current tends to cause large output offset voltages, or the input impedance becomes too low.

In this chapter you will find a solution to these problems in the form of the instrumentation amplifier (IA). All of these amplifiers are differential amplifiers, but offer superior performance over the simple DC differential amplifiers of the last chapter. The instrumentation amplifier can offer higher input impedance, higher gain, and better common-mode rejection than the single-device DC differential amplifier.

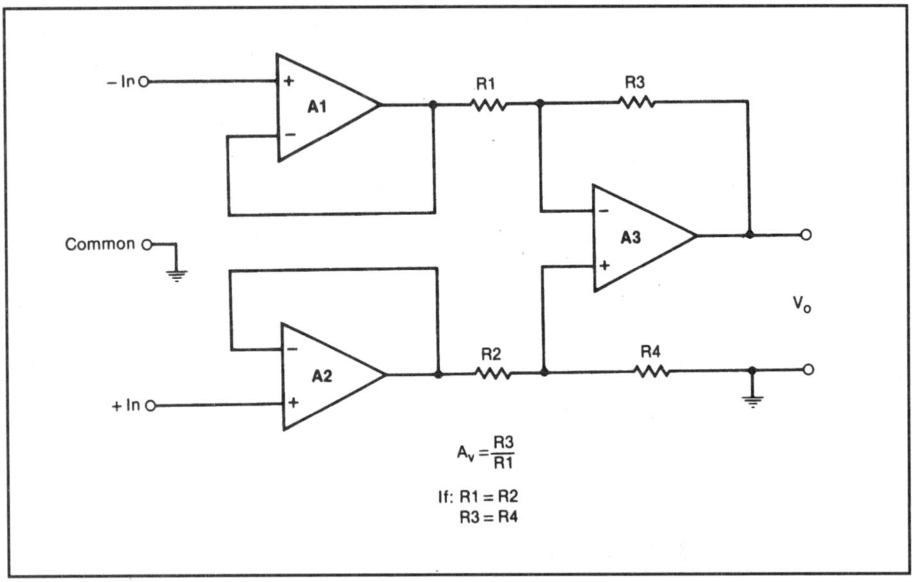

Figure 7-1. Simple DC instrumentation amplifier circuit.

Simple IA Circuit

The simplest form of instrumentation amplifier circuit is shown in Figure 7-1. In this circuit the input impedance is improved by connecting inputs of a simple DC differential amplifier (A3) to two input amplifiers (A1 and A2) that are each of the unity gain, noninverting follower configuration (their use here is as buffer amplifiers). The input amplifiers offer an extremely high input impedance (a result of the noninverting configuration) while driving the input resistors of the actual amplifying stage (A3). The overall gain of this circuit is the same as for any simple DC differential amplifier:

$$A_v = \frac{R3}{R1} \qquad \text{eq. (7-1)}$$

— where:

A_v is the voltage gain.
Assuming that R1 = R2, and R3 = R4.

It is considered best practice if A1 and A2 are identical operational amplifiers. In fact, it is advisable to use a dual operational amplifier for both A1 and A2 (i.e., two op amps in a common IC package). The common thermal environment of the dual operational amplifier will reduce thermal drift problems. The very high input impedance of "superbeta" (i.e., Darlington), BiMOS and BiFET operational amplifiers make them ideal for use as the input amplifiers in this type of circuit.

One of the biggest problems with the circuit of Figure 7-1 is that it wastes two good operational amplifiers. The most common form of instrumentation amplifier circuit uses the input amplifiers to provide voltage gain in addition to a higher input impedance. Such amplifier circuits are discussed in the next section.

Standard Instrumentation Amplifiers

The standard and most commonly-seen instrumentation amplifier is shown in Figure 7-2. Like the simple circuit discussed above, this circuit uses three operational amplifiers. The biggest difference is that the input amplifiers (A1 and A2) are now used in the "noninverting follower with gain" configuration.

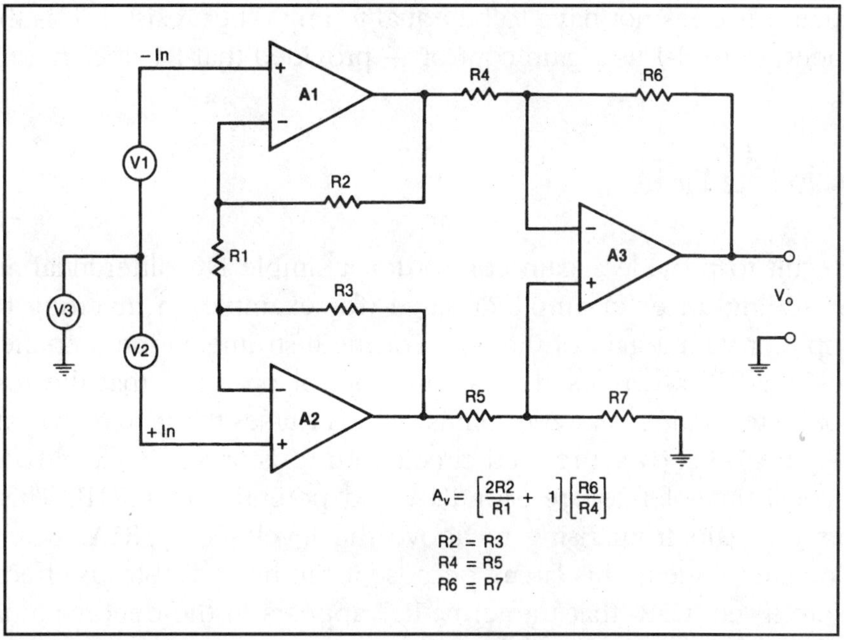

Figure 7-2. Instrumentation amplifier with more flexible gain control.

Like the circuit of Figure 7-1, the input amplifiers are ideally BiMOS, DiFET, BiFET or superbeta input types in order to obtain the maximum possible input impedance. Again, for best thermal performance use a dual, triple or quad operational amplifier for this application. The signal voltages shown in Figure 7-2 follow the standard pattern: voltages V1 and V2 form the differential input signal (V2-V1).

Gain Equation For the IA Circuit

The instrumentation amplifier consists of a pair of noninverting followers driving a simple DC differential amplifier. Not surprisingly, the gain equation for the IA is, therefore, a combination of the two:

- $$A_v = \left(\frac{2R2}{R1} + 1\right)\left(\frac{R6}{R4}\right) \qquad \textit{eq. (7-2)}$$

Provided that R2 = R3, R4 = R5 and R6 = R7.

Notice in the gain equation that resistor R1 is in the denominator. Further, notice that it is does not have to be equal to any other resistor. It is therefore convenient to use R1 as a gain control — provided that R1 does not try to go to zero.

Gain Control For the IA

It is difficult to provide a gain control for a simple DC differential amplifier without adding an extra amplifier stage (for example an inverting follower post-amplifier with a gain of 0 to -1). For the instrumentation amplifier, however, resistor R1 can be used as a gain control provided that the resistance does not go to a value near zero ohms (which causes the gain to go extremely high). Figure 7-3 shows a revised circuit with resistor R1 replaced by a series circuit consisting of fixed resistor R1A and potentiometer R1B. This circuit prevents the gain from rising to above the level set by R1A. Don't use a potentiometer alone in this circuit because it can have disastrous effect on the gain. Note in eq. (7-2) that the term "R1" appears in the denominator. If the value of R1 gets close to zero, then the gain goes very high (in fact, suppos-

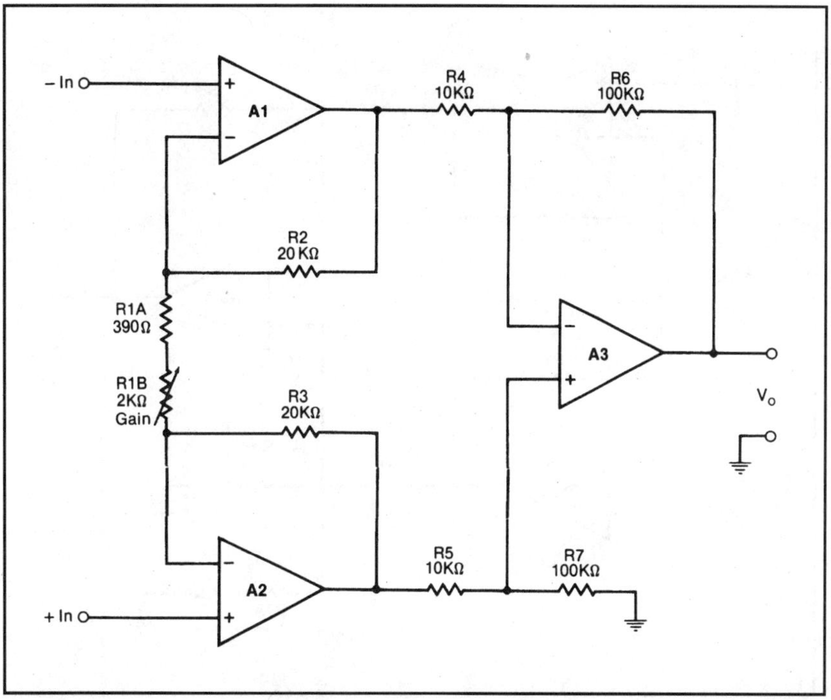

Figure 7-3. Instrumentation amplifier with variable gain control.

edly to infinity if R1 = 0). The maximum gain of the circuit should be set by using the fixed resistor in series with the potentiometer. The gain of the circuit in Figure 7-3 varies from a minimum of 167 (when R1B is set to 2000 ohms) to a maximum of 1025 (when R1B is zero). The gain expression for Figure 7-3 is:

- $$A_v = \left(\frac{2R2}{R1a+R1b}+1\right)\left(\frac{R6}{R4}\right) \qquad \text{eq. (7-3)}$$

— or, rewriting eq. (7-3) to take into account for the fact that R1A is fixed,

- $$A_v = \left(\frac{2R2}{390+R1b}+1\right)\left(\frac{R6}{R4}\right) \qquad \text{eq. (7-4)}$$

— where:

R1b varies from 0 to 2000 ohms.

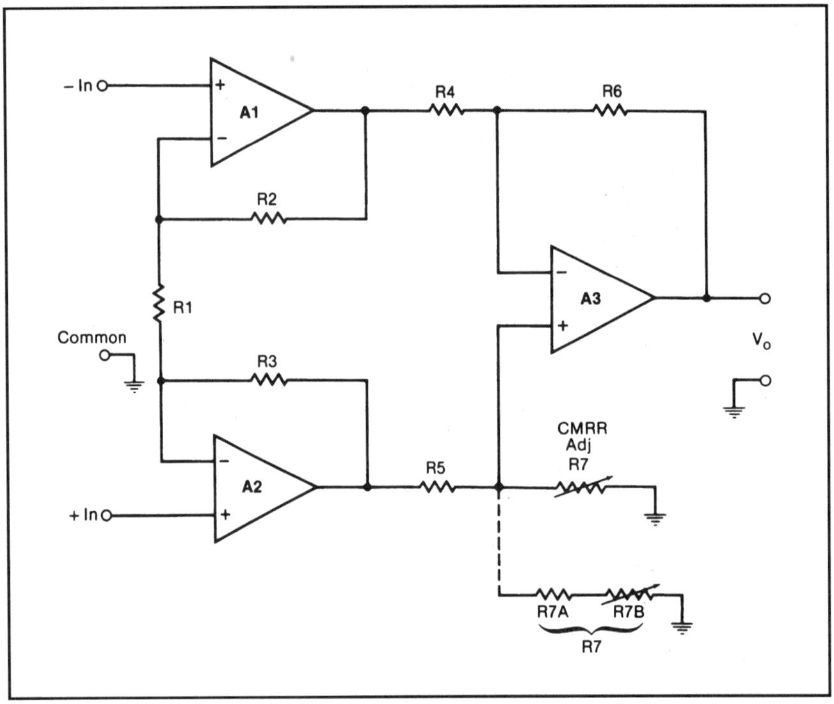

Figure 7-4. CMRR adjustment for instrumentation amplifier.

Common Mode Rejection Ratio (CMRR) Adjustment

The instrumentation amplifier is no different from any other practical DC differential amplifier in that there will be imperfect balance for common mode signals. The operational amplifiers are not ideally matched, so there will be a small gain imbalance. This gain imbalance is further deteriorated by the mismatch of the resistors. The result is that the instrumentation amplifier will respond at least to some extent to common mode signals. As in the simple DC differential amplifier circuit we can provide a common mode rejection ratio (CMRR) adjustment by making resistor R7 variable (see Figure 7-4).

One configuration of Figure 7-4 uses a single potentiometer (R7) that has a value that is ten to twenty percent larger than the required resistance of R6. For example, if R6 is 100K, then R7 should be 110 to 120K. Unfortunately, these values are somewhat difficult to obtain, so we would pick a standard value for R7 (e.g., 100K), and then select a value for R6 that is somewhat lower (e.g., 82K or 91K).

The second configuration of Figure 7-4 (shown by dotted line) uses a fixed resistor in series with a potentiometer. The general rule is to make R7A approximately 80 percent of the total required value, and R7B 40 percent of the required value. As was true in the other configuration, the sum of R7A and R7B is approximately 110 to 120 percent of the value of resistor R6. Resistance values of this sort permit the total resistance to vary from less than the nominally correct value to greater than that value, ensuring that the actually correct value is achieved.

AC Instrumentation Amplifiers

What is the principal difference between "DC amplifiers" and "AC amplifiers?" A DC amplifier will amplify both AC and DC signals up to the frequency limit of the particular circuit being used. The AC amplifier, on the other hand, will not pass or amplify DC signals. In fact, AC amplifiers will not pass AC signals of frequencies from close to DC to some lower -3 dB bandpass limit. The gain

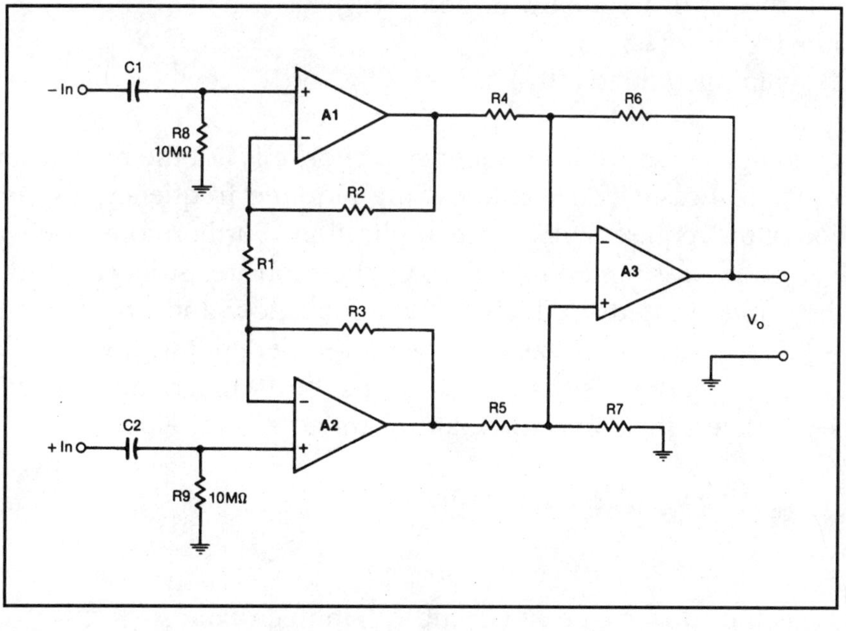

Figure 7-5. AC-coupled instrumentation amplifier.

in the region between near-DC and the "full-gain" frequencies within the passband rises at a rate determined by the design, usually +6 dB/octave (an "octave" is a 2:1 frequency change). The standard low-end point in the frequency response curve is defined as the frequency at which the gain drops off -3 dB from the full gain.

Figure 7-5 shows a modified version of the instrumentation amplifier that is designed as an AC amplifier. The input circuitry of A1 and A2 is modified by placing a capacitor in series with each op amp's noninverting input. Resistors R8 and R9 are used to keep the input bias currents of A1 and A2 from charging capacitors C1 and C2. In some modern low-input current operational amplifiers these resistors are optional because of the extremely low levels of bias current that are normally present.

The -3 dB frequency of the amplifier in Figure 7-5 is a function of the input capacitors and resistors (assuming that R8 = R9 = R and C1 = C2 = C):

$$F = \frac{1{,}000{,}000}{2\pi R C_{\mu F}} \qquad \textit{eq. (7-5)}$$

— where:

F is the -3 dB frequency in hertz (Hz)
R is in ohms (Ω)
$C_{\mu F}$ is in microfarads (µF)

The equation given above for frequency response is not the most useful form. In most practical cases you will know the required frequency response from evaluation of the requirements of the application. Furthermore, you will know the value, or at least minimum value, of the input resistors (R8 and R9) because they are selected for high input impedance, and are by convention either >10X or >100X the source impedance (depending upon the application). Typically, these resistors are selected to be 10 megohms. You will therefore need to select the capacitor values from eq. (7-6) below:

$$C_{\mu F} = \frac{1{,}000{,}000}{2\pi R F} \qquad \textit{eq. (7-6)}$$

— where:

$C_{\mu F}$ is the capacitance of C1 and C2, in microfarads (µF).

R is the resistance of R8 and R9 in ohms (Ω).
F is the -3 dB frequency in hertz (Hz).

The AC instrumentation amplifier can be adapted to all the other modifications of the basic circuit discussed earlier in this chapter. We may, for example, use a gain control (replace R1 with a fixed resistor and a potentiometer), or add a CMRR ADJ control. In fact, these adaptations are probably necessary in most practical AC IA circuits.

In many instrumentation amplifier applications it is desirable to provide selectable AC or DC coupling, as well as the ability to ground the input of the amplifiers. This latter feature is especially desirable in circuits where an oscilloscope, strip-chart paper recorder, digital data logger or computer is used to receive the data. By grounding the input of the amplifier (without also grounding

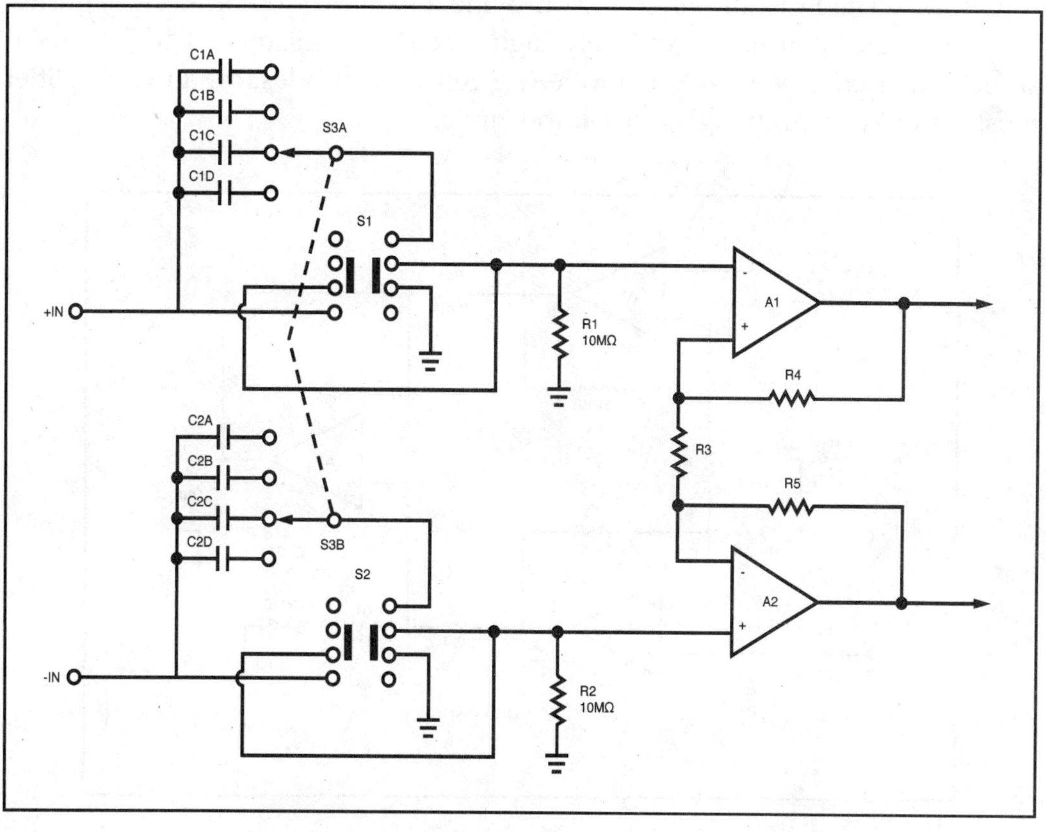

Figure 7-6. AC-GND-DC coupling input switching scheme.

the source, which could be dangerous), it is possible to set (or at least determine) the $V_d = 0$ baseline. Figure 7-6 shows a modified input circuit that uses a switch to select AC-GND-DC coupling. In addition, a second switch is provided that sets the low-end -3 dB frequency response point according to eq. (7-6) above.

An ECG Amplifier

The heart in man and animals produces a small electrical signal that can be recorded through skin surface electrodes and displayed on a cathode ray oscilloscope or paper strip-chart recorder. This electrical signal is called the electrocardiograph or ECG signal. The peak value of the ECG signal at the skin surface electrodes is on the order of one millivolt (1 mV). In order to produce a 1 volt signal to apply to a recorder or oscilloscope, therefore, a gain of 1000 is required. The ECG amplifier must provide a gain of 1000 or more. Furthermore, because skin has a relatively high electrical resistance, 1 to 20K measured with 1 cm silver-silver chloride (Ag-AgCl) electrodes, the ECG amplifier must also have a very high input impedance.

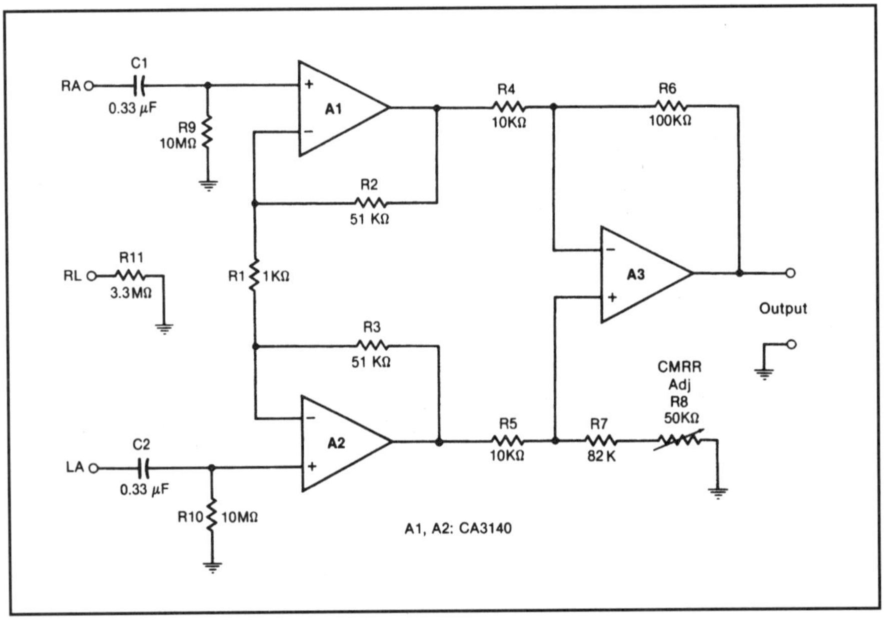

Figure 7-7. Animal ECG amplifier based on the IA circuit (do not use on humans).

Another requirement for the ECG amplifier is that it be an AC amplifier. The reason for this requirement is that metallic electrodes (such as Ag-AgCl) applied to the electrolytic skin acts like a small battery and produces a DC "half-cell potential." This potential tends to be on the order of 1 to 2 volts, so it is more than 1000 times higher than the signal voltage. By making the amplifier respond only to AC, we eliminate the artifact caused by the DC half-cell potential.

The frequency selected for the -3 dB point of the ECG amplifier must be very low, close to DC, because the standard ECG waveform contains very low frequency components. The typical ECG signal has significant frequency components in the range 0.05 to 100 hertz, which is the industry standard frequency response for diagnostic ECG amplifiers (some clinical monitoring ECG amplifiers use 0.05 Hz to 40 Hz to eliminate muscle artifact due to patient movements).

The typical ECG amplifier has differential inputs because the most useful ECG signals are differential in nature, and in order to suppress 60 Hz hum picked up on the leads and the patient's body. In the most simple case, the right arm (RA) and left arm (LA) electrodes form the inputs to the amplifier, with the right leg (RL) defined as the common (see Figure 7-7). The basic configuration of the amplifier in Figure 7-7 is the AC-coupled instrumentation amplifier discussed earlier. The gain for this amplifier is set to slightly more than X1000, so a one-millivolt ECG peak signal will produce a 1-volt output from this amplifier. Because of the high gain it is essential that the amplifier be well balanced. This requirement suggests the use of a dual amplifier for A1 and A2. An example might be the CA-3240 device, which is a dual BiMOS device that is essentially two CA-3140s in a single eight-pin miniDIP package. Also in the interest of balance, 1 percent or less tolerance resistors should be used for the equal pairs.

The lower-end -3 dB frequency response point is set by the input resistors and capacitors. In this case, the particular combination of values shown produces a response of 0.048 Hz.

The CMRR ADJUST control in Figure 7-7 is usually a ten- to twenty-turn trimmer potentiometer. It is adjusted in the following manner.

1. Short together the RA, LA and RL inputs.

2. Connect a DC voltmeter to the output (either a digital voltmeter or an analog meter with a 1.5 volt DC scale; alternatively, a DC-coupled oscilloscope can be used, but be sure to identify the zero baseline.

3. Adjust CMRR ADJUST control (R7) for zero volts output. Disconnect the RL input, and connect a signal generator between RL and the still-connected RA/LA terminal.

4. Adjust the output of the signal generator for a sinewave frequency in the range 10 to 40 Hz, and a peak-to-peak potential of 1 volt to 3 volts.

5. Using an AC scale on the voltmeter (or an oscilloscope), again adjust CMRR ADJUST (R7) for the smallest possible output signal. It may be necessary to readjust the voltmeter or oscilloscope input range control to greater sensitivity in order to observe the best null.

6. Remove the RA/LA short. The ECG amplifier is ready to use.

A suitable post amplifier for the ECG preamplifier and other IA circuits is shown in Figure 7-8. This amplifier is placed in the signal line between the output of Figure 7-7 and the input of the oscilloscope or paper chart recorder used to display the waveform. The gain of the post amplifier is variable from 0 to +2; it will produce a 2 volt maximum output when a one millivolt ECG signal provides a 1 volt output from the preamplifier. Because of the high level signals used, this amplifier can use ordinary 741 operational amplifiers.

The frequency response of the amplifier is set to an upper -3 dB point of 100 Hz, with the response dropping off at a -6 dB/octave rate above that frequency. This frequency response point is determined by capacitor C3 operating with resistor R12:

- $$F_{Hz} = \frac{1{,}000{,}000}{2\pi RC} \qquad eq.\ (7\text{-}7)$$

- $$F_{Hz} = \frac{1{,}000{,}000}{(2)(3.14)\left(10^5\right)(0.015\mu F)} \qquad eq.\ (7\text{-}8)$$

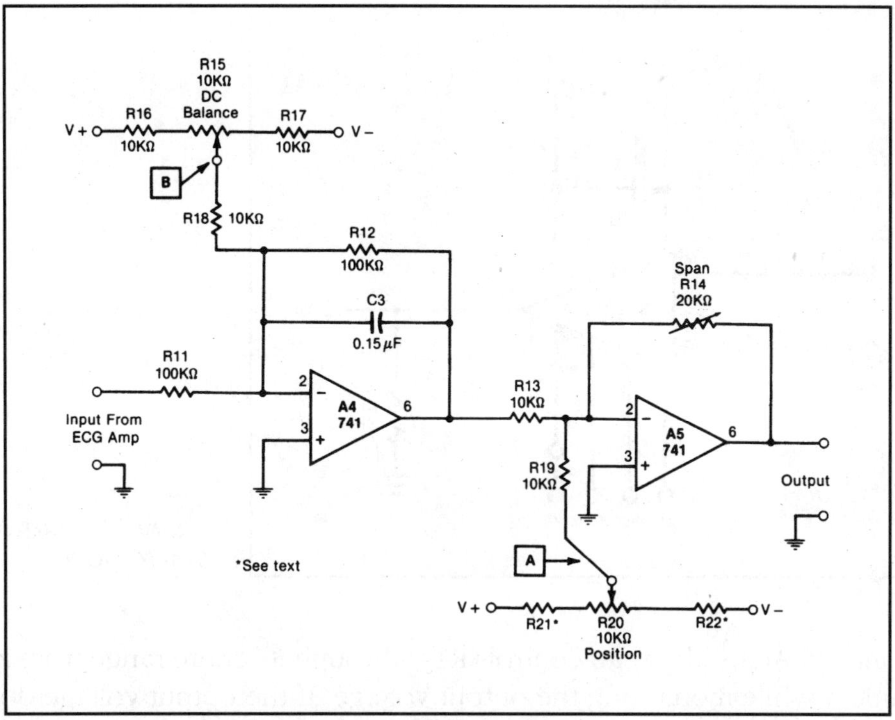

Figure 7-8. Post-amplifier for ECG preamplifier.

There are three controls in the post amplifier circuit: span, position and DC balance. The span control is the 0 to 2 gain control, and the label "span" reflects instrumentation users' language, rather than electronics language. The position control sets the position of the output waveform on the display device. Resistors R21 and R22 are selected to limit the travel of the beam or pen to full scale. Set these resistors so that the maximum potential at the end terminals of R20 corresponds to full-scale deflection of the display device.

The DC balance control is used to cancel the collective effects of offset potentials created by the various stages of amplification. This control is adjusted as follows:

Follow the CMRR adjust procedure, and then reconnect the short-circuit at RA, LA and RL. Move the voltmeter to the output of Figure 7-8. Adjust the position control for zero volts at point "A." Adjust the DC balance control for zero volts

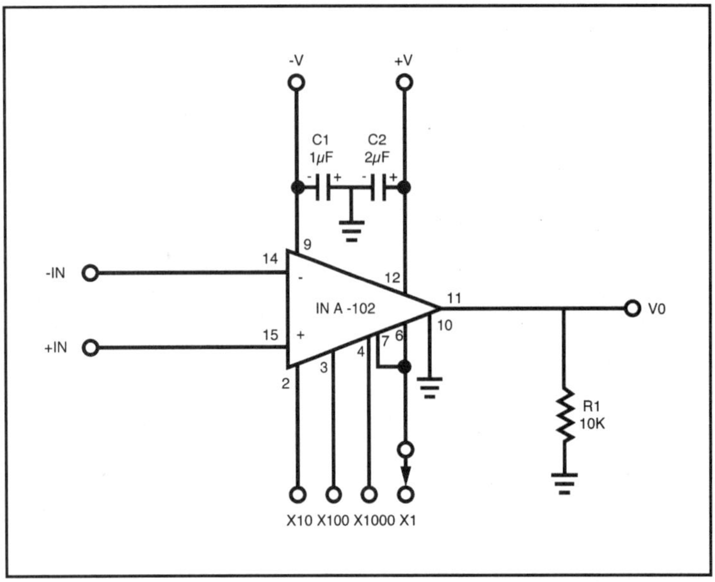

Figure 7-9. Burr-Brown INA-102 ICIA.

at point "B." Adjust the span control (R14) through its entire range from zero to maximum while monitoring the output voltage. If the output voltage does not shift, then no further adjustment is needed.

If the output voltage in the previous step varied as the span control is varied, then adjust the DC balance until varying the span control over its full range does not produce an output voltage shift. Repeat this step several times until no further improvement is possible.

Remove the RA, LA, RL short; the amplifier is ready for use.

Some Commercial IC IA Examples

Now that we've seen the basic circuitry of the instrumentation amplifier, and how to build them from discrete operational amplifiers, let's take a look at a couple of commercial integrated circuit IA examples: the Burr-Brown INA-102 and INA-117 devices.

The INA-102 is shown in Figure 7-9. It is a monolithic ICIA in a 16-pin DIP package. Four internal switch-selectable voltage gain levels are available: X1,

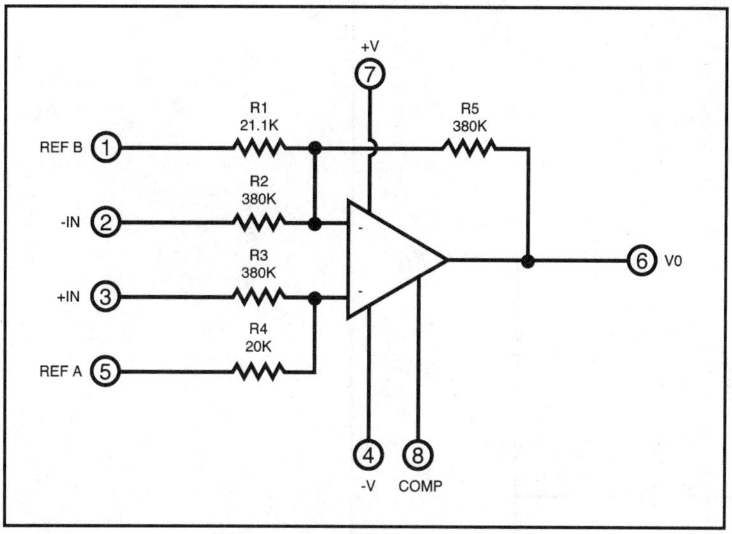

Figure 7-10. Burr-Brown INA-117 ICIA.

X10, X100 and X1000. The INA-102 offers at least 90 dB CMRR, and a low drift characteristic. It is intended for the amplification of low-level signals from sources such as strain gauges, thermocouples and piezoresistive bridges.

The INA-117 device is shown in Figure 7-10. The resistors shown in the figure are internal to the INA-117. Programming gain is achieved by the external connections to these resistors. The INA-117 is designed to accommodate a high common-mode input voltage range (±200 volts), and is input-protected to ±500 volts. The device is basically unit gain, but can be used with a post-amplifier to boost gains to any range desired.

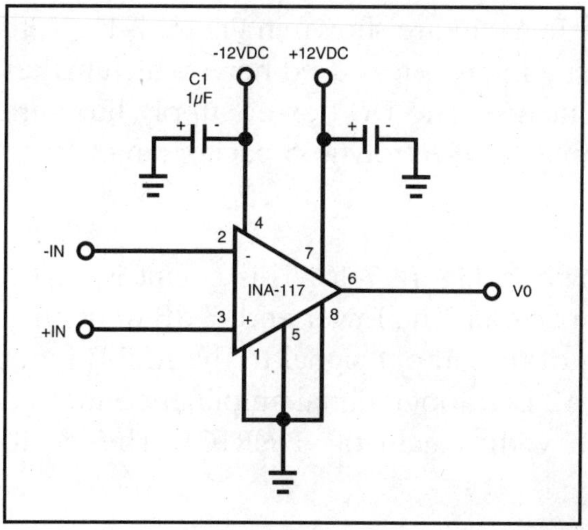

Figure 7-11. INA-117 circuit connections.

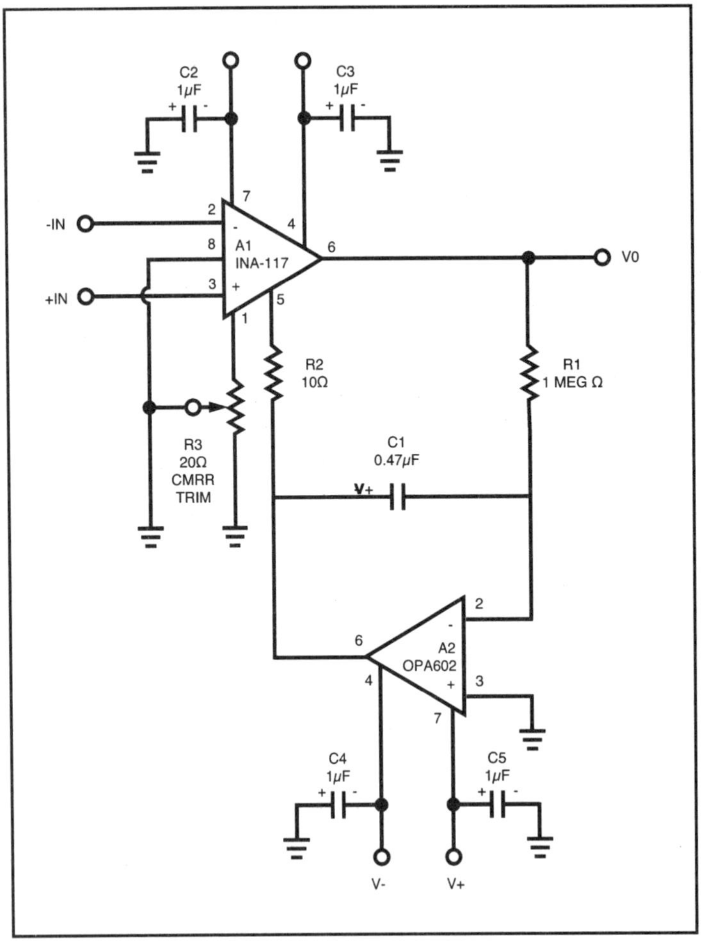

Figure 7-12. "Capacitorless" AC coupling using feedback.

The basic external connections for the INA-117 are shown in Figure 7-11. Note that there are no external gain programming resistors used here, which makes the device relatively easy to use in practice. The DC power supply lines are each bypassed to ground with 1 µF tantalum electrolytic capacitors in order to preserve stability.

An application of the INA-117 is shown in Figure 7-12. This circuit is an AC coupled IA without input coupling capacitors. The lower end -3 dB frequency limit is set by feeding back a portion of the output signal to the REF-B input (pin no. 5) of the INA-117. Amplifier A2 is an operational amplifier connected in the Miller integrator configuration with a gain of $-1/\omega R1C1$. The -3 dB frequency (F_{co}) is set by:

- $$F_{co} = \frac{GAIN}{2\pi R1C1}$$ *eq. (7-9)*

When R1 = 1 megohm and C1 = 0.47 μF, the cut-off frequency is about 7 Hz, so this circuit can be used for applications where a desired signal is riding on an undesired DC voltage level.

Next Chapter

In the next chapter we will take a look at isolation amplifiers. These amplifiers are used whenever an extremely high impedance is needed between the inputs of the amplifier and the DC power supplies, or the AC power mains that are used to provide power to the overall circuit.

Chapter 8

Isolation Amplifiers

There are a number of situations where ordinary amplifiers are either in danger themselves because of the electrical environment, or present a danger to the users (e.g., in medical equipment). An example of the former might be an amplifier in a high voltage experiment such as an electrophoresis system, while the latter is represented by cardiac monitors and other devices used in the hospital.

An isolation amplifier is one in which there is an extremely high impedance (e.g., 10^{12} Ω) between the signal inputs and those power supply terminals that are connected to a DC power supply that is, in turn, connected to the AC power mains. In the case of the medical equipment, we are trying to prevent minute leakage currents from the AC power lines from being applied to the patient. Current levels that are negligible to people normally can be fatal to a hospitalized patient in some cases. In other cases, the high impedance is used to prevent high voltages at the signal inputs from adversely affecting the rest of the circuitry. Modern isolation amplifiers can provide 10^{12} ohms or more of electrical isolation between the power lines and the signal inputs.

There are several different popular circuit symbols for the isolation amplifier, but the one that seems to be most common is shown in Figure 8-1. It consists of the regular triangular amplifier symbol, but broken in the middle to indicate isolation between the "A" and "B" portions. In most cases, there will be the following connections to the isolation amplifier:

Non-Isolated "A" Side: V+ and V- DC power supply lines (to be connected to a DC supply powered by the AC lines), output to the rest of the (non-isolated) circuitry, and (in some designs) a non-isolated ground or common. This ground is connected to the chassis or main system ground also served by the main DC power supplies.

Isolated "B" Side: Isolated V+ and V-, isolated ground or common and the signal inputs. The isolated power supply and ground are not connected to the main power supply or ground systems. In some cases, batteries are used for the isolated side, while in others special DC power supplies derived from the main supplies are used (of which, more later).

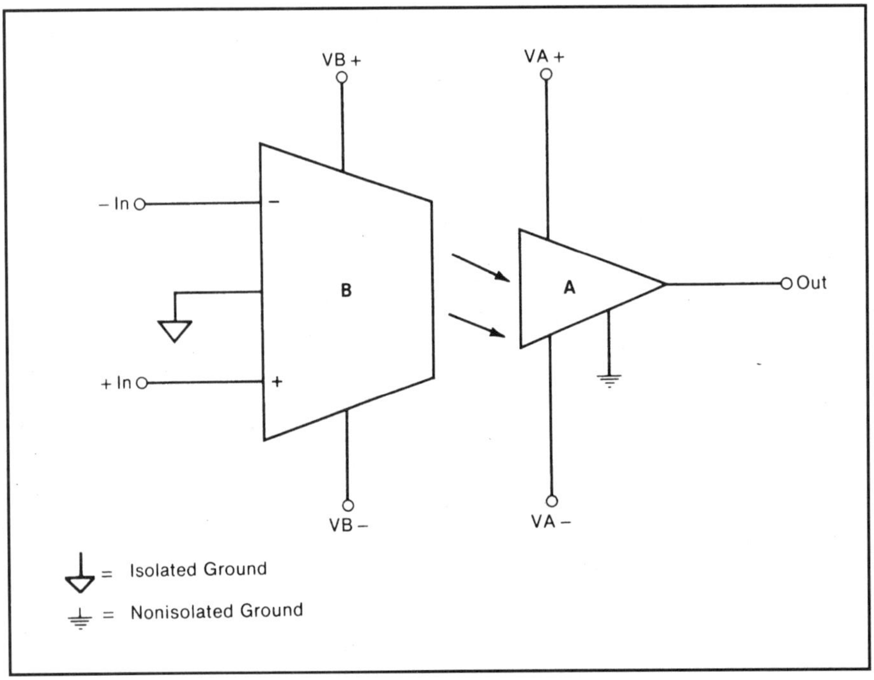

Figure 8-1. Symbol for isolation amplifier.

Different manufacturers use different approaches to the design of isolation amplifiers: battery-powered, carrier-operated, optically-coupled, and current loading.

Battery-Powered Designs

This approach is perhaps the simplest to implement, but it is not always most suitable for your convenience due to problems inherent in battery upkeep. A few products exist, however, that use a battery-powered front-end amplifier, even though the remainder of the product is AC-powered.

Carrier-Operated

Figure 8-2 shows an isolation amplifier that uses the carrier technique. The circuitry inside of the dashed line is isolated from the AC power lines (in other words, the "B" side of Figure 8-1). In most cases, the voltage gain of the isolated section is in the medium gain range of X10 to X500.

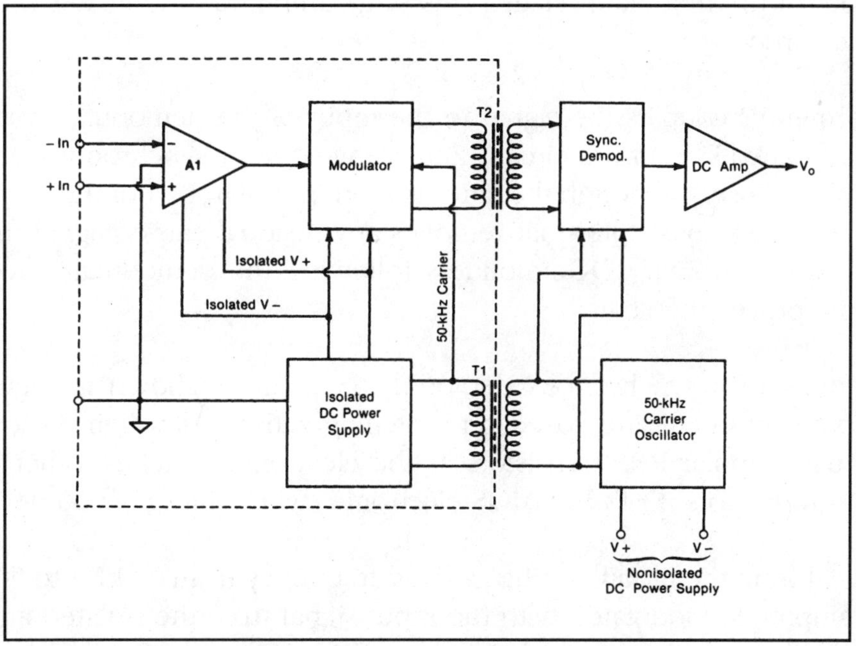

Figure 8-2. Transformer-coupled carrier-type isolation amplifier.

The isolation is provided by separation of the ground, power supply and signal paths into two mutually exclusive sections by ultrasonic transformers T1 and T2. These transformers have a design and core material that works very well in the ultrasonic (20 kHz to 500 kHz) region, but is terribly inefficient at the 60 Hz frequency used by the AC power lines. This design feature allows the transformers to easily pass the high frequency carrier signal, while attenuating severely all 60 Hz energy that might be present.

Although most models use a carrier frequency in the 50 to 60 kHz range, it is possible to find examples of carrier amplifier in the entire 20 to 500 kHz spectrum.

The carrier oscillator signal is coupled through transformer T1 to the isolated stages. Part of the energy from the secondary of T1 is directed to the modulator stage; the remainder of the energy is rectified and filtered, and then used as an isolated DC power supply. The DC output of this power supply is used to power the input "B" amplifiers and the modulator stage.

An analog signal applied to the input is amplified by A1, and is then applied to one input of the modulator stage. This stage amplitude modulates the signal onto the carrier.

Transformer T2 couples the signal to the input of the demodulator stage on the non-isolated side of the circuit. Either envelope or synchronous demodulation may be used, although the latter is more popular. Part of the demodulator stage is a low-pass filter that removes any residual carrier signal from the output signal. Ordinary DC amplifiers following the demodulator complete the signal processing chain.

An example of a synchronous demodulator circuit is shown in Figure 8-3. These types of circuit are based on switching action. Although the example shown uses bipolar PNP transistors as the electronic switches, other circuits use NPN transistors, FETs or CMOS electronic switches (e.g., 4066 device).

The signal from the modulator has a fixed frequency from 20 kHz to 500 kHz, and is amplitude-modulated with the input signal from the isolated amplifier. This signal is applied to the emitters of transistors Q1 and Q2 (via T1) in push-

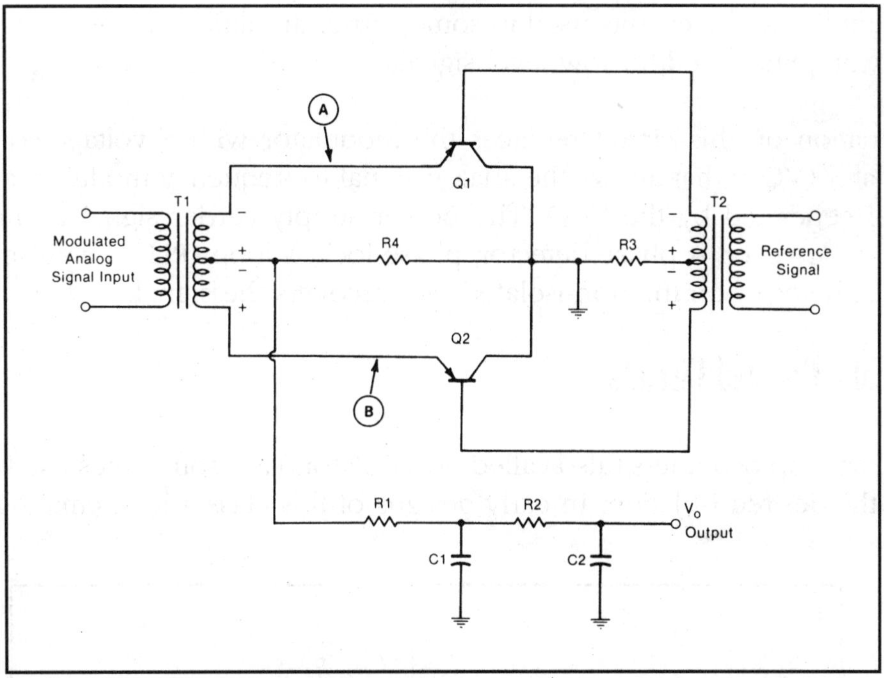

Figure 8-3. Synchronous demodulator circuit.

pull. On one half of the cycle, therefore, the emitter of Q1 will be positive with respect to the emitter of Q2. On alternate half-cycles, the opposite situation occurs: Q2 is positive with respect to Q1.

The bases of Q1 and Q2 are also driven in push-pull, but by the carrier signal (called here the "reference signal"). This action causes transistors Q1 and Q2 to switch on and off, but out of phase with each other.

On one half of the cycle, we will have the polarities shown in Figure 8-3. Transistor Q1 is turned on. In this condition point "A" on T1 is grounded. The voltage developed across load resistor R4 is positive with respect to ground.

On the alternate half-cycle, Q2 is turned on, so point "B" is grounded. However, the polarities have reversed, so the polarity of the voltage developed across R4 is still positive. This causes a full-wave output waveform across R4, which when filtered becomes a DC voltage level proportional to the amplitude of the input signal. This same description of synchronous demodulators also applies to the circuits used in some carrier amplifiers (a specialized laboratory amplifier used for low-level signals).

A variation on this circuit replaces the modulator with a voltage-controlled oscillator (VCO) that allows the analog signal to frequency modulate a carrier signal generated by the VCO. The power supply carrier signal is stilled required, however. A phase detector, phase-locked loop (PLL), or pulse-counting FM detector on the non-isolated side recovers the signal.

Optically-Coupled Circuits

Electronic optocouplers (also called optoisolators) are sometimes used to provide the desired isolation. In early designs of this class a light-emitting diode

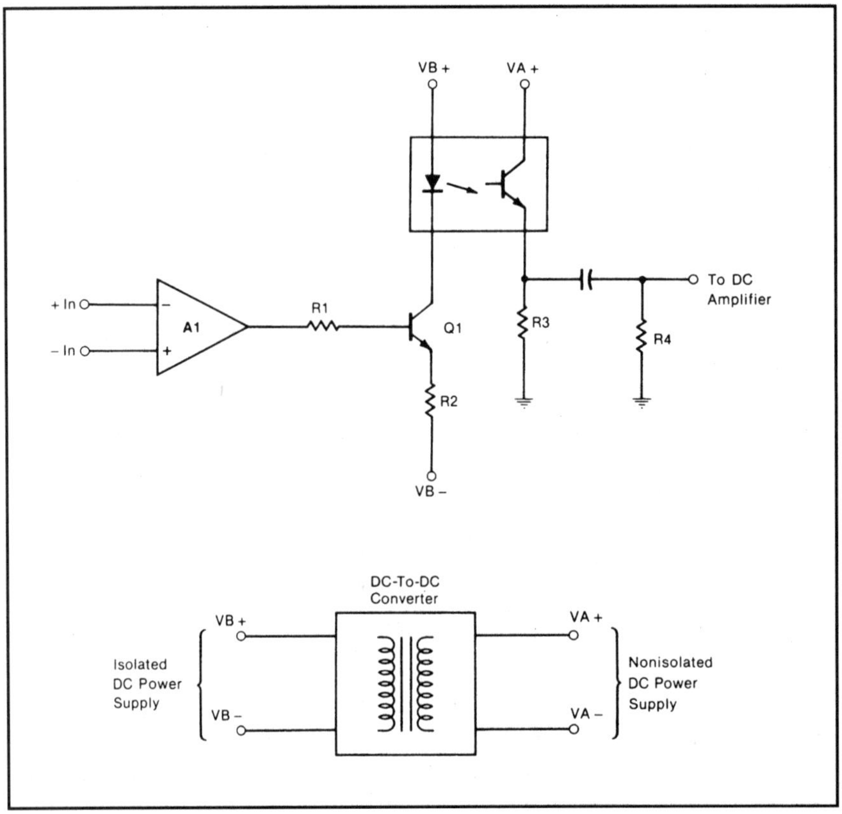

Figure 8-4. Optoisolator type isolation amplifier circuit.

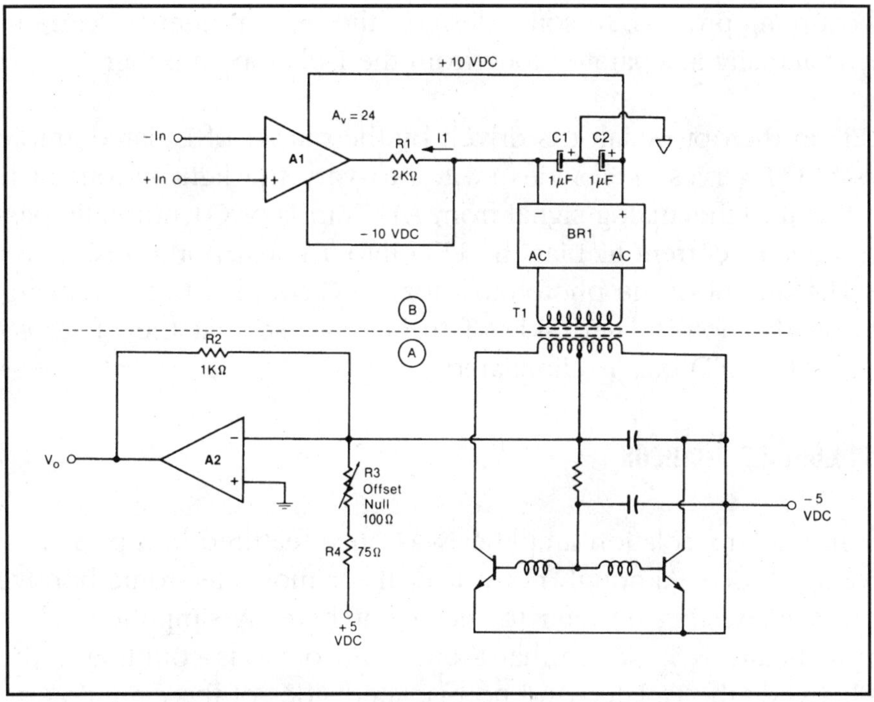

Figure 8-5. Current-starved isolation amplifier.

(LED) was sandwiched with a photoresistor or phototransistor. Modern designs, however, use integrated circuit (IC) optoisolators that contain an LED and phototransistor inside of a single DIP IC package.

There are actually several approaches to optical coupling. Two very popular methods are the carrier and direct methods. The carrier method is the same as discussed in the previous section, except that an optoisolator replaces transformer T2.

The carrier method is not the most widespread in optically-coupled isolation amplifiers because of the frequency response limitations of some IC optoisolators. Only recently have these problems been resolved.

A more common approach is shown in Figure 8-4. This circuit uses the same DC-to-DC converter to power the isolated stages as was used in other designs. This will keep A1 isolated from the AC power mains but is not used in the signal coupling process. In some designs, the high frequency "carrier" power supply is actually a separate block from the isolation amplifier.

The LED in the optoisolator is driven by the output of isolated amplifier A1. Transistor Q1 serves as a series switch to vary the light output of the LED proportional to the analog signal from A1. Transistor Q1 normally passes sufficient collector current to bias the LED into a linear portion of its operating curve. The output of the phototransistor is AC-coupled to the remaining amplifiers on the non-isolated side of the circuit, so that the offset condition created by the LED bias is eliminated.

Current Loading Methods

A current loading isolation amplifier was once featured in a popular medical monitoring device. Although not practical for most electronic hobbyists, the method is interesting enough to include it here. A simplified schematic is shown in Figure 8-5. Notice that there is no obvious coupling path for the signal between the isolated and non-isolated sides of the circuit!

The gain-of-24 isolated input preamplifier (A1) in Figure 8-5 consists of a JFET input operational amplifier, in order to provide a very high input impedance. The output of A1 is connected to the isolated -10 volt DC power supply through load resistor R1. This power supply is a DC-to-DC converter operating at 250 kHz. Transformer T1 provides isolation between the floating power supplies on the isolated "B" side of the circuit and the non-isolated "A" side of the circuit (which are AC-line powered).

An input signal causes the output of A1 to vary the current loading of the floating -10 volt DC power supply. Changing the loading proportional to the analog input signal causes variation of the T1 primary current that is also proportional to the analog signal. This current variation is converted to a voltage variation by amplifier A2. An offset null control (R3) is provided in the A2 circuit to eliminate the offset at the output due to the quiescent current flowing when the analog input signal is zero. In that case, the current loading of T1 is constant — but still provides an offset to the A2 amplifier.

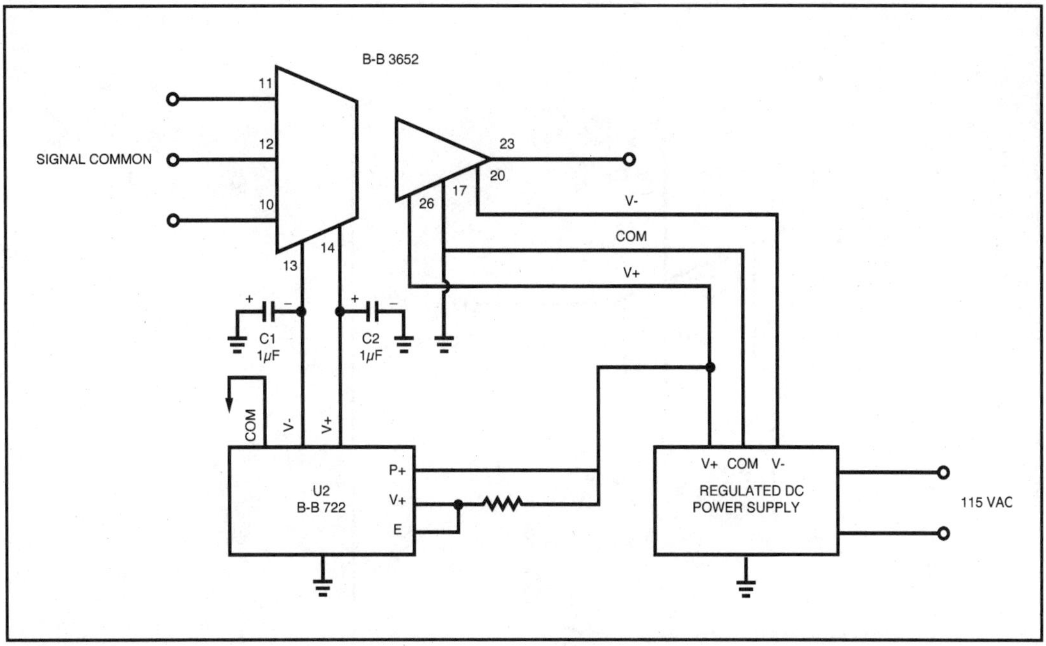

Figure 8-6. Burr-Brown 3652 isolation amplifier with isolated DC power supply provided by B-B 722 device.

Some Commercial Examples

In this section we take a look at a couple commercial examples of actual isolation amplifiers.

Burr-Brown 3652

Figure 8-6 shows the circuit of an isolation amplifier based on the Burr-Brown 3652 isolation amplifier device. The DC power for both the isolated and non-isolated sections of the 3652 is provided by the 722 dual DC-to-DC converter. This device produces two independent ±15 VDC supplies that are each isolated from the 60 Hz AC power mains and from each other. The 722 device is powered from a +12 VDC source that is derived from the AC power mains. In some cases, the non-isolated section (which is connected to the output terminal) is powered from a bipolar DC power supply that is derived from the 60

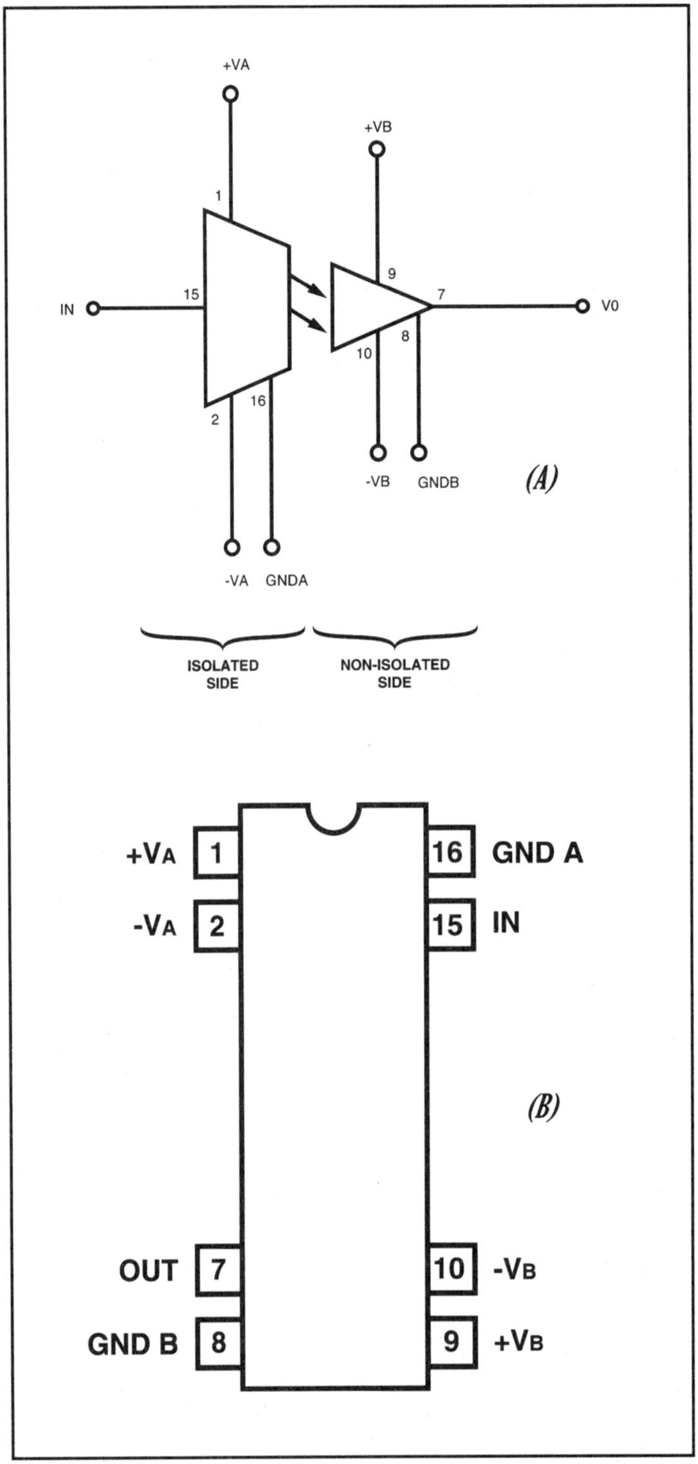

Figure 8-7. Burr-Brown ISO-122 isolation amplifier: A) pin-outs; B) package.

Hz AC mains, such as a ±12 VDC or ±15 VDC supply. In no instance, however, should the isolated DC power supplies be derived from the AC power mains.

There are two separate ground systems in this circuit, symbolized by the small triangle and the regular three-bar "chassis" ground symbol. The isolated ground is not connected to either the DC power supply ground/common, or the chassis ground. It is kept floating at all times, and becomes the signal common for the input signal source.

The gain of the circuit is approximately:

$$A_v = \frac{1{,}000{,}000 \; ohms}{R1+R2+115 \; ohms} \qquad eq.\ (8\text{-}1)$$

In most design cases, the issue is the unknown values of the gain setting resistors. We can rearrange the equation above to solve for (R1 + R2):

$$(R1+R2) = \frac{1{,}000{,}000-(115A_v)}{A_v} ohms \qquad eq.\ (8\text{-}2)$$

— where:

R1 and R2 are in ohms (Ω).

A_v is the voltage gain desired.

Burr-Brown ISO122P

The ISO122P device is an isolation amplifier built in a 16-pin single-wide plastic DIP package. One of the endearing features of the ISO122P is that it provides isolation amplifier function at a low cost. According to a recent catalog, the price in unit quantities is less than $20. The ISO122P features unity gain, a high isolation mode rejection of 140 dB, and 0.020% non-linearity. It can be used alone, or in conjunction with other circuitry on either the isolated or non-isolated sides. Figure 8-7A shows the schematic symbol with pin-outs designated, while Figure 8-7B shows the 16-pin DIP package (also with pin-out designations shown).

Figure 8-8 shows a sample application circuit in which the unity gain ISO122P is used as an output buffer between a DC differential amplifier and the rest of the circuit. The DC differential amplifier is shown here as an example only,

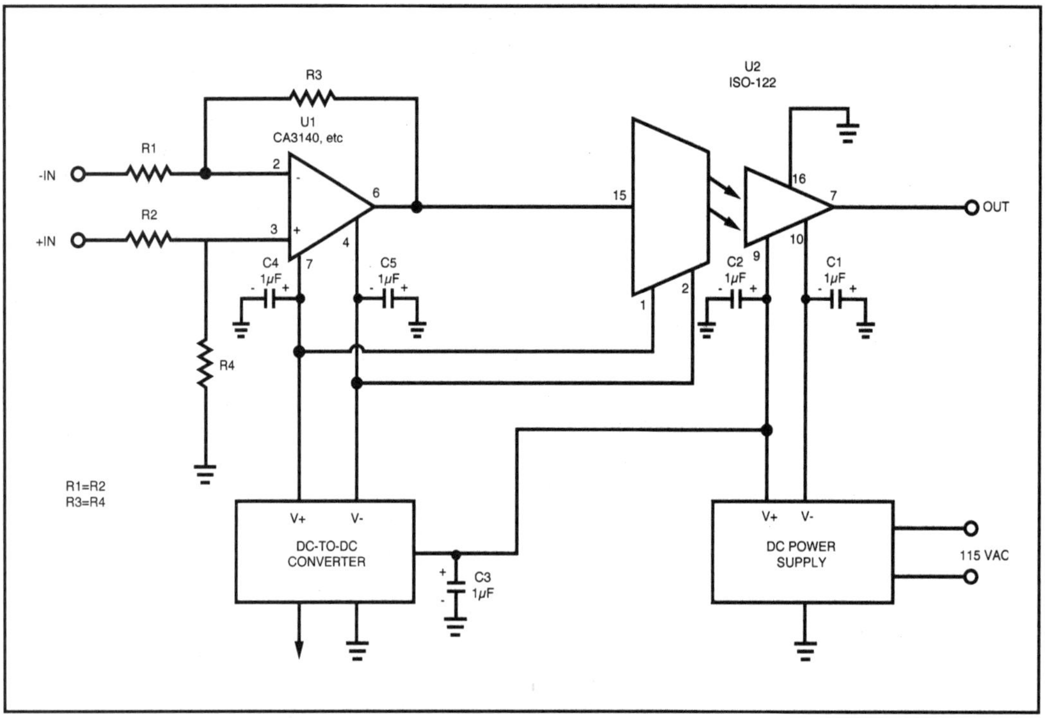

Figure 8-8. Isolated DC differential amplifier using the B-B ISO-122 device for isolation.

and can be replaced with any circuit of your choice provided that the ±10 volt input voltage range of the ISO122P is not exceeded.

If the input circuit of Figure 8-8 is used, then the operational amplifier at U1 can be a 741, CA-3140 or any other that serves the purpose. The pin-outs shown for U1 are "industry standard" for a wide variety of devices, including the 741 and CA-3140 devices. If R1 = R2, and R3 = R4, then the differential voltage gain of the circuit of Figure 8-8 is given by:

- $$A_{vd} = \frac{R3}{R1} = \frac{R4}{R2} \qquad \text{eq. (8-3)}$$

The power for the isolated side of the ISO122P is shown in the form of a DC-to-DC converter in Figure 8-8 and 8-9, of which the B-B 722 shown earlier is but one example. Alternatively, one could use batteries for the isolated side, and then ensure that they are charged through some circuit, such as an electromechanical relay, that does not destroy the isolation feature.

Conclusion

Although the isolation amplifier is considerably more expensive than common IC linear amplifiers, there are applications where these amplifiers are absolutely critical. Wherever the instrument could cause injury to a human, or wherever the environment is such that the electronics must be isolated for its own sake, the isolation amplifier is the device of choice (at least in the front end).

The isolation amplifiers are generally the solution of choice for most engineers who must interface sensors under conditions that could potentially be dangerous. In addition, they are also the solution of choice for the case where the amplifier input circuits are sometimes subjected to high voltages, either transient or steady-state.

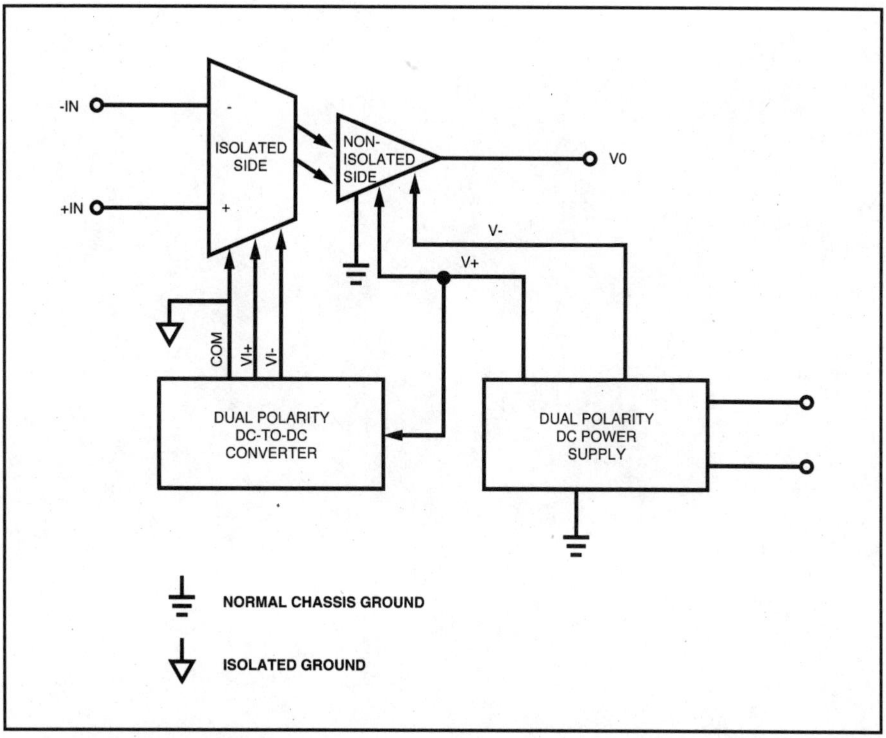

Figure 8-9. Generic power connections for isolation amplifiers.

Chapter 9

Operational Transconductance Amplifiers

The operational amplifier is a simple voltage amplifier with a transfer function of V_o/V_{in}. While that type of device is typically the most commonly used form, there are cases where we need a transconductance amplifier. That form of amplifier is based on a transfer function that relates an output current to an input voltage. In other words:

- $$G_m = \frac{I_o}{V_{in}} \qquad \textit{eq. (9-1)}$$

— where:

G_m is the transconductance in siemens.
I_o is the output current in amperes (A).
V_{in} is the input signal in volts (V).

The operational transconductance amplifier (OTA) equivalent circuit is shown in Figure 9-1. The differential input circuit is much like the input circuit of the operational amplifier because each are differential voltage inputs. The output side of the circuit, however, is a current source that produces an output current, I_o, that is proportional to the gain and the input voltage. The current gain

(A_i) of this circuit is a function of the transconductance (I_o/V_{in}) and the load resistance, R:

- $$A_i = G_m R$$ *eq. (9-2)*

— where:

A_i is the current gain (dimensionless).
G_m is the transconductance (I_o/V_{in}) in siemens (S).
R is the load resistance (one-half the output resistance R_o) in ohms (Ω).

Note: G_m and R must be expressed in equivalent reciprocal units. In other words, when G_m is in siemens then R is in ohms. Likewise, millisiemens/milliohms and microsiemens/micro-ohms are also paired. Perhaps the most common version of the operational transconductance amplifier (OTA) is the CA-3080/CA-3080A device. This popular device is available in the eight-pin metal IC package using the pin-outs shown in Figure 9-2. The CA-3080 devices will operate over DC power supply voltages from ±2 volts to ±15 volts, with adjustable power consumption of 10 microwatts to 30 milliwatts. The gain is 0 to G_mR. The input voltage spread is ±5 volts. The bias current can be set to as high as 2 mA.

Note that the pin-outs for the CA-3080 device are "industry standard" operational amplifier pin-outs, except for the bias current applied to pin number 5:

V- on pin 4.
V+ on pin 7
Inverting input on pin 2.
Noninverting input on pin 3.
Output on pin 6.

The parameters of the operational transconductance amplifier are set by the bias current (I_{bias}). For example, the transconductance is 19.2 times higher than the bias current:

- $$G_m = 19.2I$$ *eq. (9-3)*

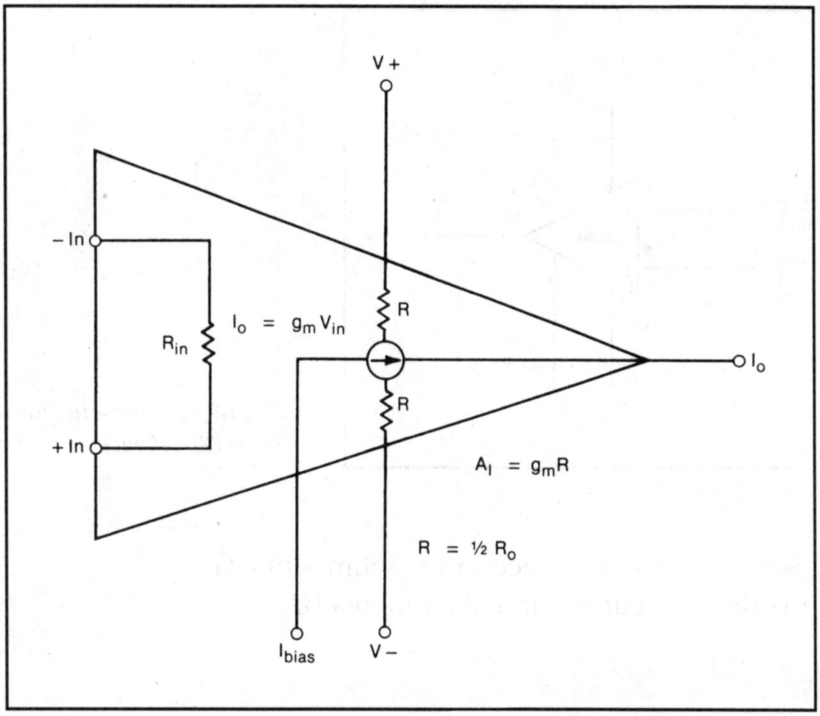

Figure 9-1. Equivalent circuit for an operational transconductance amplifier (OTA).

— where:

G_m is in millisiemens.

I_{bias} is in milliamperes.

In many actual design cases we will know the required value of G_m from our knowledge of I_o/V_{in}, so can set the G_m required by adjusting the bias current. In those cases, the I_{bias} is found by rewriting the above expression:

$$I_{bias} = \frac{G_m}{19.2} \qquad \text{eq. (9-4)}$$

The output resistance of the device is also a function of the bias current:

$$R_o = \frac{7.5}{I_{bias}} \qquad \text{eq. (9-5)}$$

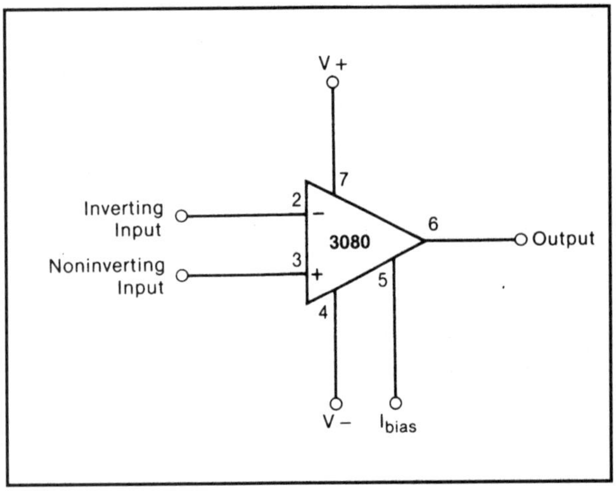

Figure 9-2. Pin-outs for the CA-3080 OTA device.

— where:

R_o is the output resistance in megohms (megΩ).

I_{bias} is the bias current in milliamperes (mA).

Example

What is the output resistance (R_o) when the bias current is 500 uA (i.e., 0.5 mA)?

Ro = 7.5/I_{bias}.

Ro = 7.5/0.5.

Ro = 15 megohms.

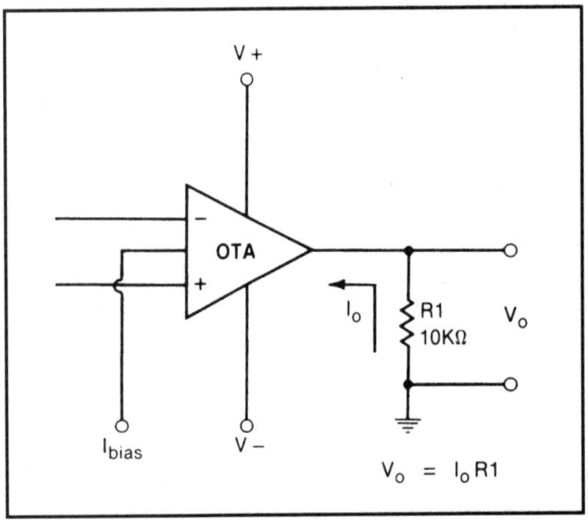

Figure 9-3. Converting the OTA output current to a voltage signal.

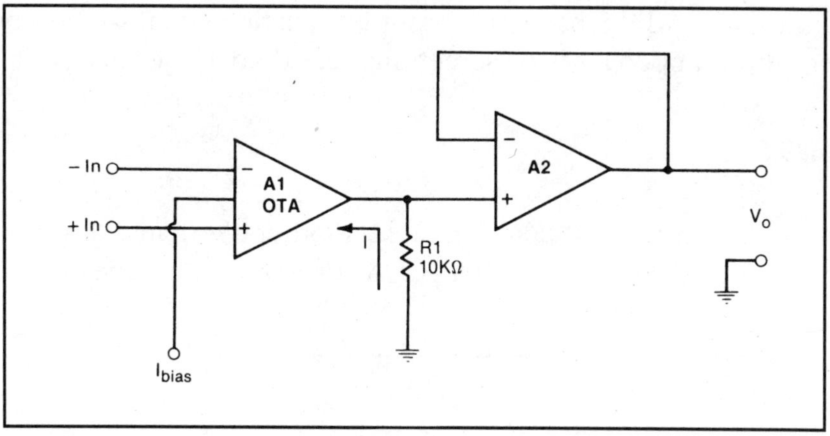

Figure 9-4. Low output impedance current-to-voltage output converter.

Voltage Amplifier From the OTA

The OTA is a current-output device, but can be used as a voltage amplifier when one of the following circuit strategies are used. The simplest method is the resistor load shown in Figure 9-3. Because the output of the OTA is a current (I_o), we can pass this current through a resistor (R1) to create a voltage drop. The value of the voltage drop (and the output voltage, V_o) is found from Ohm's law:

- $$V_o = I_o R1$$ *eq. (9-6)*

A problem with this circuit is that the source impedance is very high, being equal to the value of R1. In the example shown in Figure 9-3, the output impedance is 10K. We can overcome this problem by adding a unity gain noninverting operational amplifier such as A2 in Figure 9-4. The output voltage in this case is the same as for the non-amplifier version: I_o R1. Another form of low-impedance output circuit is shown in Figure 9-5. This form uses the inverting follower configuration of the operational amplifier (A2). The output voltage is the product of the OTA output current (I_o) and the op amp feedback resistor (R1):

- $$V_o = I_o R1$$ *eq. (9-7)*

Both Figures 9-4 and 9-5 have an output impedance equal to the operational amplifier output impedance, or something less than 100 ohms.

Next Chapter

In the next chapter we will take a look at another alternative linear IC amplifier, the current difference amplifier (CDA) or Norton amplifier.

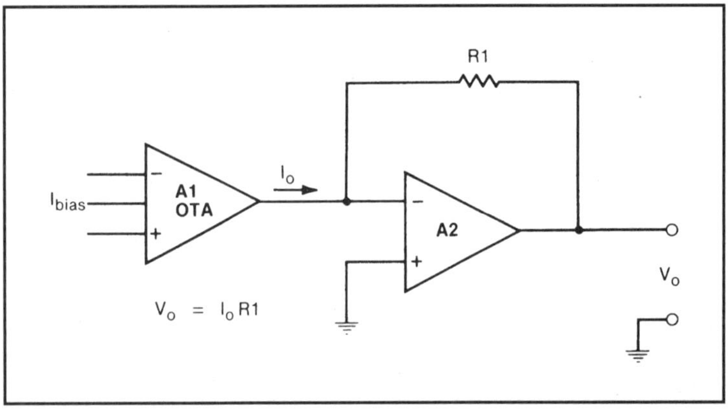

Figure 9-5. Alternate low impedance output circuit.

Chapter 10

Current Difference Amplifiers

The current difference amplifier (CDA), also called the Norton amplifier (after Norton's current theorem), is a non-operational amplifier that performs in much the same way as the op amp, but not quite exactly the same. The CDA has certain features that make it uniquely useful for certain applications. One place where the CDA is more useful than the operational amplifier is in circuits that process AC signals, but are limited to a single polarity DC power supply. For example, in automotive equipment we are limited to a single 12 to 14.4 volt DC battery power supply that uses the car chassis for negative common return.

There are other cases where the linear IC amplifier is but a minor feature of the circuit, most of which operate from a single DC power supply. It would wasteful in such circuits to use the operational amplifier. We would either have to bias the operational amplifier with an external resistor network, or provide a second DC power supply.

The normal circuit symbol for the CDA is shown in Figure 10-1. This symbol looks much like the regular op amp symbol, except that a "current source" is placed along the side opposite the apex (the circle with an arrow enclosed).

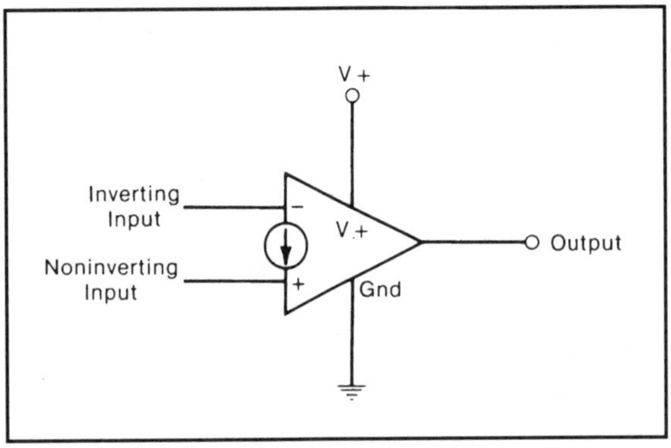

Figure 10-1. Normal circuit symbol for current difference amplifier.

This symbol is typically used for several products such as the National Semiconductor LM-3900 device, which is a quad Norton amplifier. You may sometimes find schematics where the op amp symbol is used for the CDA, but that is technically incorrect usage.

CDA Configuration

The input circuit of the CDA differs radically from the operational amplifier. Recall that the op amp used a differential amplifier driven with a constant current source (CCS) supplying the collector-emitter current. The CDA is quite different, however, as can be seen in Figure 10-2.

The overall circuit of a typical CDA is shown in Figure 10-2A, while an alternate form of the input circuit is shown in Figure 10-2B. Transistor Q7 in Figure 10-2A forms the output transistor, while Q5 is the driver. Both the NPN output transistor and the PNP driver transistor operate in the emitter follower configuration. Transistors Q4, Q5, Q6 and Q8 are designed as current sources. The input transistor is Q3, and it operates in the common emitter configuration. The base of Q3 forms the inverting (-IN) input for the CDA.

The noninverting input of the CDA is formed with a "current mirror" transistor, Q1 (transistor Q1 in Figure 10-2A is "diode connected" and serves exactly the same function as diode D1 in Figure 10-2B). The dynamic resistance offered by the current mirror transistor (Q2) is given by:

$$r = \frac{26}{I_b} \qquad \text{eq. (10-1)}$$

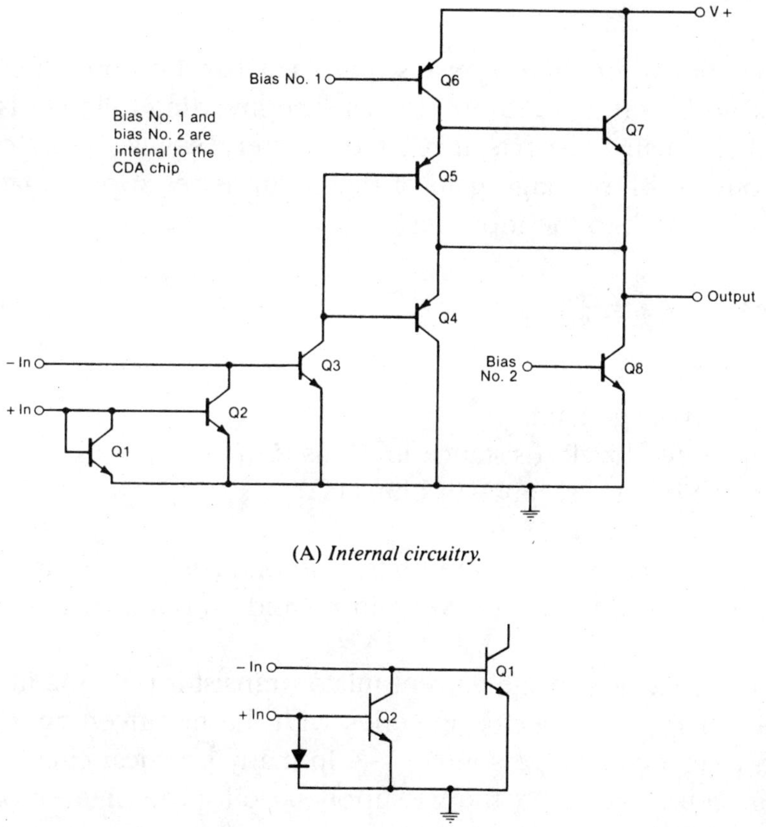

(A) *Internal circuitry.*

(B) *Current mirror input circuit.*

Figure 10-2. Overall circuit for a typical CDA.

— where:

r is the dynamic resistance of Q2 in ohms (Ω).

I_b is the base bias current of Q3 in milliamperes (mA).

Equation (10-1) is used only at normal room temperature, and will vary with wide temperature excursions. For most readers, however, the room temperature version of the equation will suffice. Data sheets for specific current difference amplifiers give additional details for amplifiers that must operate outside of the relatively narrow temperature range specified by the equation.

Inverting Follower Circuits

Like its operational amplifier cousins, the CDA can be configured in either inverting or noninverting follower circuits. The inverting follower is shown in Figure 10-3. In many respects this circuit is very similar to its operational amplifier cousins. The voltage gain of the circuit is set approximately by the ratio of the feedback to the input resistor:

- $A_v = \frac{-R2}{R1}$ *eq. (10-2)*

— where:

A_v is the voltage gain.
R2 is the feedback resistance in ohms (Ω).
R1 is the input resistance in ohms (Ω).

(Note: R1 and R2 are in the same units. The minus sign indicates that a 180 degree phase reversal occurs between input and output signals.)

We must provide a bias to the current mirror transistor (i.e., Q2 in Figure 10-2B), so resistor R_{ref} is connected in series with the noninverting input of the CDA and a reference voltage source, V_{ref}. In many practical circuits the reference voltage source is merely the V+ supply used for the CDA. In other cases, however, some other potential might be required, or alternatively the reference current is required to be regulated more tightly (or with less noise) than the supply voltage. Ordinarily we set the reference current to some convenient value between 5 μA and 100 μA. For V+ power supply values of +12 VDC, for example, it is common to find a 1 megohm resistor used for R_{ref}. In that case, the input reference current is: I_{ref} = 12/1,000,000 = 12 μA.

A constraint placed on CDAs is that the input resistor (R1) used to set gain must be high compared with the value of the current mirror dynamic resistance (eq. [10-1] above). The CDA becomes nonlinear (i.e., distorts the input signal) if the input resistor value approaches the current mirror resistance, r. In that case, the voltage gain is not -R2/R1, but rather:

- $A_v = \frac{R2}{R1+r}$ *eq. (10-3)*

— where:

A_v is the voltage gain.
R2 is the feedback resistance in ohms (Ω).
R1 is the input resistance in ohms (Ω).
r is the current mirror resistance (Ω).

Equation (10-3) essentially reduces to eq. (10-2) when we can force R1 to be very much larger than r. This goal is easily achieved in most circuits because r is tiny (work a few examples with normal bias currents in eq. [10-1].).

The output voltage of the CDA will exhibit an offset potential even when the AC input signal is zero. This potential is given by:

$$V_o = \left(\frac{V_{ref}R2}{R_{ref}}\right) + \left(1 - \frac{R2}{R_{ref}}\right)\phi \qquad \text{eq. (10-4)}$$

— where:

Vo is the output potential in volts (V).
Vref is the reference potential in volts (usually V+).
R2 is the feedback resistance in ohms (Ω).
Rref is the current mirror bias resistance in ohms (Ω).
ϕ is a temperature dependent factor (≅0.7 volts for room temperature).

The capacitor in series with the input circuitry has the effect of limiting the low-end frequency response. The -3 dB cut-off frequency is a function of the value of this capacitor and the input resistance, R1. This frequency, F, is given by:

$$F = \frac{1{,}000{,}000}{2\pi R1 C1_{\mu F}} \qquad \text{eq. (10-5)}$$

— where:

F is the lower-end -3 dB frequency in hertz (Hz).
R1 is in ohms (Ω).
C1 is in microfarads (μF).

In some CDA circuits there is also a capacitor in series with the output terminal. The purpose of that capacitor is to prevent the DC offset that is inherent in this type of circuit from affecting following circuits. The output capacitor will

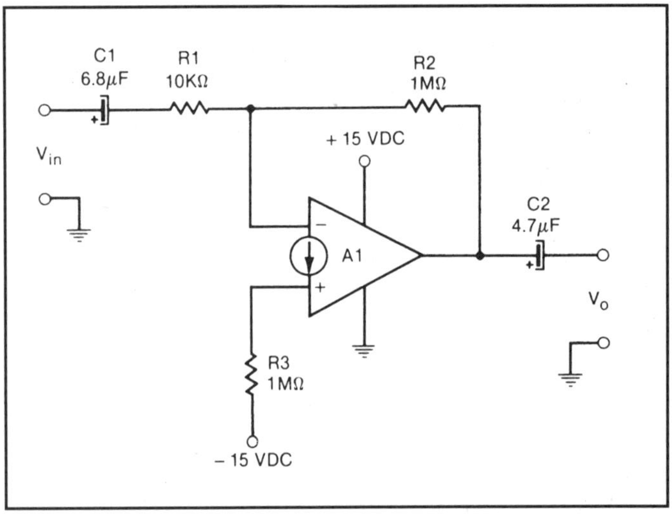

Figure 10-3. Inverting follower CDA circuit.

also limit the low end frequency response. The same form of equation (i.e., eq. [10-5]) is used to determine this frequency, but using the input resistance of the load as the "R" term.

As is often the case with equations presented in electronic books, eq. (10-5) is not in the most useful format. In most cases, we will know the input resistance (R1) from the application. It is typically not less than 10 times the source impedance, and forms part of the gain equation. Consideration of driving

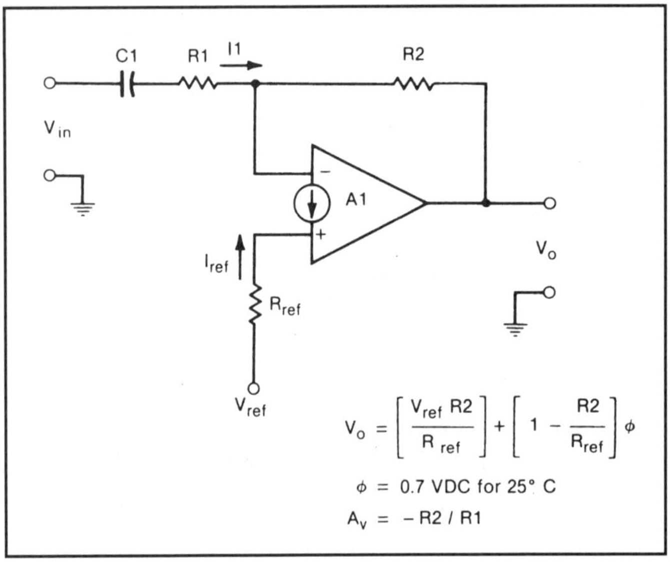

Figure 10-4. Gain-of-100 CDA circuit.

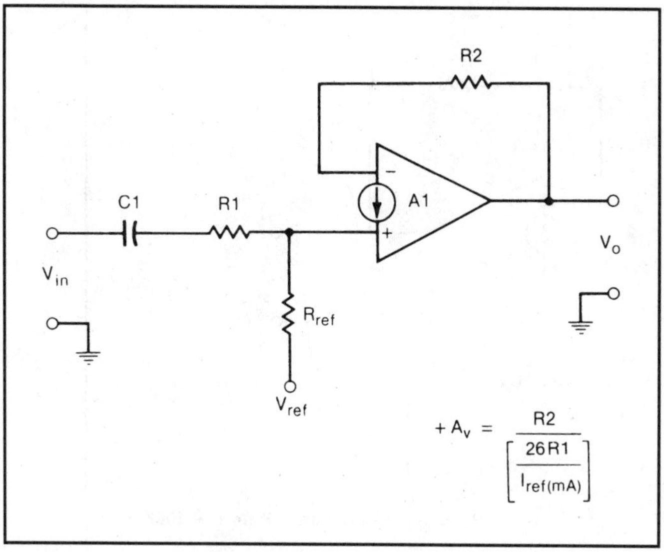

Figure 10-5. Noninverting follower CDA circuit.

source impedance and voltage gain tends to determine the value of R1. The required low-end frequency response is usually determined from the application. We generally know (or can find out) that frequency spectrum of input signals. From the lower limit of the frequency spectrum we can determine F. Thus, we will determine F and R1 from other considerations than the circuit. We therefore need a version of eq. (10-5) that will allow us to calculate capacitor C1:

$$C1_{uF} = \frac{1{,}000{,}000}{2\pi R1F} \qquad \textit{eq. (10-6)}$$

(All terms as defined for eq. [10-5].)

Designer's Casebook Example

Design a gain-of-100 AC amplifier based on a CDA in which the input impedance is at least 10 kohms and the -3 dB frequency response is 3 Hz or lower. Assume DC power supplies of ±15 volts DC. (See Figure 10-4 for final circuit.)

1. Set the reference current to the noninverting input to a value between 5 and 100 μA. Select 15 μA.

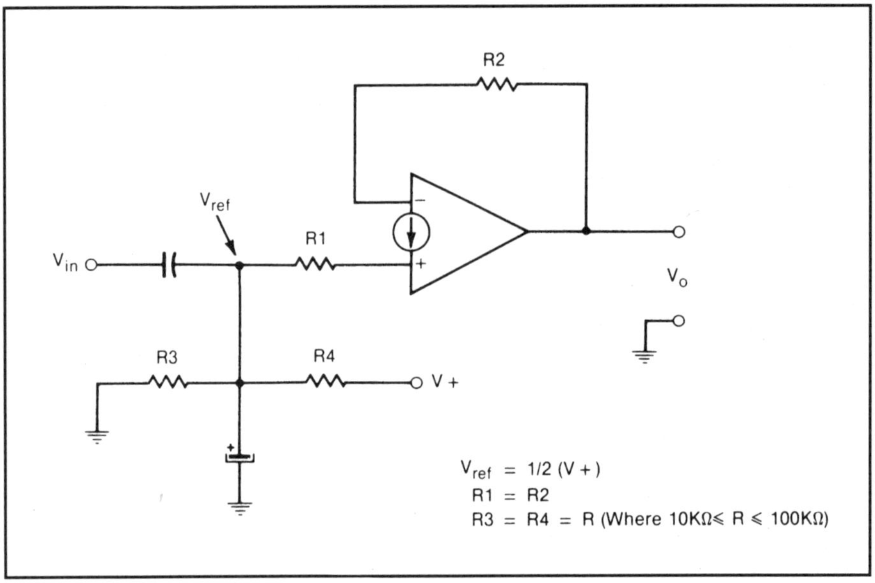

Figure 10-6. Single DC supply operation of the CDA.

R3 = V/I_{ref}
R3 = (15 VDC)/(0.000015)
R3 = 1 megohm

2. Set the gain resistors. R1 can be 10,000 ohms in order to meet the input impedance requirement.

(From Av = R2/R1, we know that R2 = AvR1)

R2 = AvR1
R2 = (100) x (10,000)
R2 = 1,000,000 ohms = 1 megohm

3. Find the value of input capacitor C1 when F(-3 dB) is 3 hertz.

- $$C1_{\mu F} = \frac{1,000,000}{2\pi R1F}$$ *eq. (10-7)*

- $$C1_{\mu F} = \frac{1,000,000}{(2)(3.14)(10,000\Omega)(3Hz)} = 5.3\mu F$$ *eq. (10-8)*

Because 5.3 µF is a nonstandard value, we select the next higher standard value (which is 6.8 µF).

The value of output capacitor C2 can be arbitrarily set to 6.8 µF if the load impedance is 10,000 ohms. If the load impedance is higher than that value, then use 4.7 µF or 6.8 µF. If the load impedance is very much higher than 10,000 ohms, or is lower than 10,000 ohms, then calculate the value using the same equation as for C1 but with the load resistance substituted for the input impedance.

Noninverting Amplifier Circuits

The noninverting amplifier CDA configuration is shown in Figure 10-5. This circuit retains the reference current bias applied to the noninverting input, but rearranges some of the other components. As in the case of the inverting amplifier configuration, the noninverting amplifier uses R2 to provide negative feedback between the output terminal and the inverting input. Unlike the inverting CDA circuit, however, input resistor R1 is connected in series with the noninverting input. The gain of the noninverting CDA amplifier is given by:

- $$+A_v = \frac{R2}{\left(\frac{26R1}{I_{ref(mA)}}\right)} \qquad \text{eq. (10-9)}$$

— where:

+Av is the voltage gain.
R1 and R2 are in ohms (Ω).
I_{ref} is the bias current in milliamperes (mA)

The reference current, I_{ref}, is set to a value between 5 µA and 100 µA (i.e., 0.005 to 0.1 mA). Unlike the situation in the inverting amplifier, the value of this current is partially responsible for setting the gain of the circuit. Some clever designers have even used this current as a limited gain control for some CDA stages. The value of the resistor that provides the reference current (R_{ref}) is set by Ohm's law, considering the required value of reference current and the reference voltage, V_{ref}. In most common applications, the reference voltage is merely one of the supply voltages. The value of R_{ref} is determined from:

$$R_{ref} = \frac{V_{ref}}{I_{ref}} \qquad \text{eq. (10-10)}$$

— where:

R_{ref} is the reference resistor in ohms (Ω).
V_{ref} is the reference potential in volts.
I_{ref} is the reference current in amperes (note: 1 μA = 0.000001 ampere).

The value of input impedance is approximately equal to R1 if R1 is much higher than the dynamic resistance of the current mirror inside the CDA (which is typically the case).

As was true in the inverting follower case, the input capacitor (C1) sets the low-end frequency response of the amplifier. The -3 dB frequency is given by exactly the same equation as for the inverting case (see eqs. [10-5] and [10-6]).

Figure 10-6 shows a modification of the noninverting follower circuit that allows for a noisy reference source. This type of amplifier circuit might be used where the DC power supplies that are used for the reference voltage are electrically noisy. Such noise could come from other stages in the circuit, or from outside sources.

The purpose of Figure 10-6 is to form a reference voltage from a resistor voltage divider circuit consisting of R3 and R4. The value of V_{ref} will be:

$$V_{ref} = \frac{(V+)R3}{R3+R4} \qquad \text{eq. (10-11)}$$

— where:

V_{ref} is the reference potential in volts (V).
V+ is the supply potential in volts (V).
R3 are in ohms (Ω).

Inspection of eq. (10-11) reveals that V_{ref} = (V+)/2 when R3 = R4, which is the usual case in practical circuits. The reference current is:

$$I_{ref} = \frac{V_{ref}}{R1} \qquad \text{eq. (10-12)}$$

Super-Gain Amplifier

There is a practical limit to voltage gain using standard resistor values and standard circuit configurations (a similar problem also exists for operational amplifiers). In Figure 10-7 we see a means for overcoming this limitation. This super-gain amplifier forms a noninverting follower in a manner similar to the earlier circuit, except that feedback resistor R2 is driven from an output voltage divider network rather than directly from the output terminal of the CDA. The voltage gain of the circuit of Figure 10-7 is given by:

$$A_v = \left(\frac{R2}{R1}\right)\left(\frac{R3+R4}{R3}\right) \qquad \text{eq. (10-13)}$$

Example

Find the gain of the circuit shown in Figure 10-7 if R1 = 10K, R2 = 100K, R3 = 10K and R4 = 100K.

$$A_v = \left(\frac{R2}{R1}\right)\left(\frac{R3+}{R}\right. \qquad \text{eq. (10-14)}$$

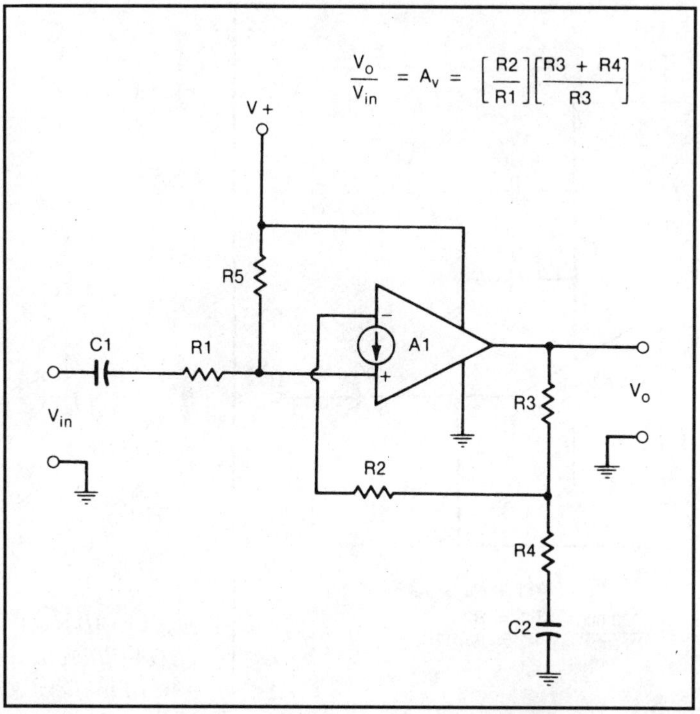

Figure 10-7. Single supply AC amplifier based on CDA device.

- $$A_v = \left(\frac{100K}{10K}\right)\left(\frac{100K+10K}{10K}\right)$$ *eq. (10-15)*

- $$A_v = (10)(11) = 110$$ *eq. (10-16)*

Capacitor C1 is set using the same eq. (10-5) that was used previously, while C2 is set to have a capacitive reactance of R4/10 at the lowest frequency of operation (in other words, the low end -3 dB point).

Differential Amplifiers

A differential amplifier is one that will produce an output that is proportional to the gain and the difference between potentials applied to the inverting and noninverting inputs. Figure 10-8 shows the circuit of a CDA differential amplifier. It is similar to the operational amplifier version of this simple circuit in several respects. For example, the two input resistors are equal and the differential voltage gain is the ratio of the negative feedback resistor (R3) and the input resistor:

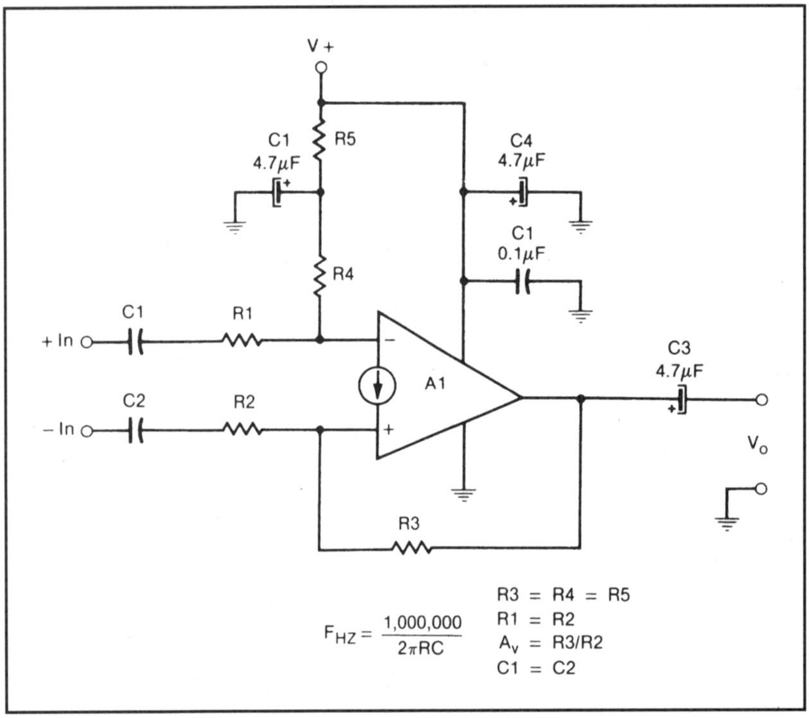

Figure 10-8. A CDA differential AC amplifier.

$$A_v = \frac{R3}{R2}$$ eq. (10-17)

If: R1 = R2 and R3 = R4 = R5

The differential input impedance of this circuit is twice the value of the input resistances: R_{in} = R1 + R2. The bias current is provided through the two series resistors, R4 and R5.

Assuming that R1 = R2 = R, and C1 = C2 = C, we can calculate the low-end -3 dB frequency from the equation:

$$F = \frac{1,000,000}{2\pi RC} Hz$$ eq. (10-18)

— where:

F is the -3 dB frequency in hertz (Hz).
R is in ohms (Ω).
C is in microfarads (μF).

When this circuit is used as a 600-ohm line receiver, we can make the two input resistors 330 ohms each (or 270 ohms) if a small mismatch can be tolerated. Ideally, the input resistors will be 300 ohms each (which may require two resistors for each input resistor).

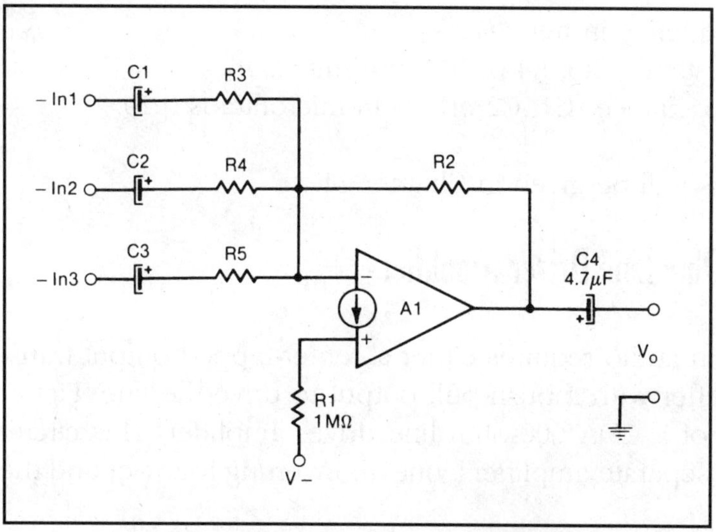

Figure 10-9. CDA summation amplifier can be used as an audio mixer.

AC Mixer/Summer Circuits

The CDA mixer or summer circuit is shown in Figure 10-9. This circuit is used to combine two or more inputs into one channel. The basic circuit is an inverting follower. Each input sees a gain that is the quotient of the feedback resistor to its input resistor:

- $$A_v(1) = \frac{-R2}{R3}$$ *eq. (10-19)*
- $$A_2(2) = \frac{-R2}{R4}$$ *eq. (10-20)*
- $$A_v(3) = \frac{-R2}{R5}$$ *eq. (10-21)*

From eqs. (10-19) through (10-21) we can deduce that the output voltage is found from:

- $$V_o = R2\left(\frac{V1}{R3} + \frac{V2}{R4} + \frac{V3}{R5}\right)$$ *eq. (10-22)*

The frequency response of each channel is found from the usual equation for -3 dB frequency:

- $$F = \frac{1{,}000{,}000}{2\pi RC}$$ *eq. (10-23)*

— where:

F is the -3 dB frequency in hertz (Hz).
R is the input resistance (R3, R4 or R5) in ohms (Ω).
C is the input capacitance (C1, C2 or C3) in microfarads (μF).

More about audio mixers will be given in Chapter 11.

Differential Output 600-Ohm Line Driver Amplifier

The 600-ohm line used in audio requires either a center-tapped output transformer, or a linear amplifier with a push-pull output to drive the line. Figure 10-10 shows the circuit of a CDA 600-ohm line driver amplifier. This circuit basically consists of two separate amplifiers, one an inverting follower and the

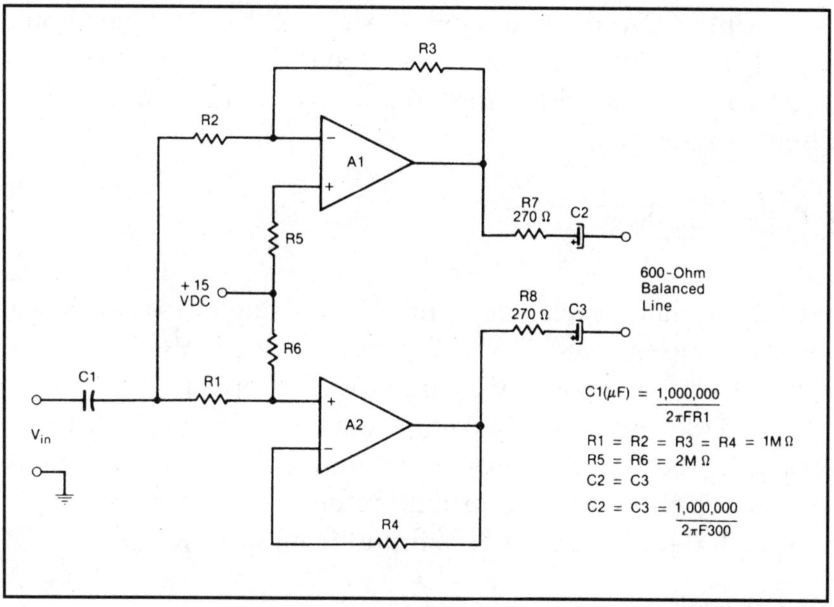

Figure 10-10. Line driver amplifier circuit.

other a noninverting follower. The bias resistors (R5 and R6) are set to provide a small bias current in the 5- to 100-μA range. This current is found from Ohm's law, $I_{ref} = (V+)/R$. In the example shown in Figure 10-10 the resistors are set to 2 megohms for a supply voltage of +15 volts DC.

The capacitors in the circuit set the low-end -3 dB point in the frequency response curve. These capacitor values are set from the following (assuming that R1 = R2 = R3 = R4, and R5 = R6):

- $$C1 = \frac{1{,}000{,}000}{2\pi FR1}$$ *eq. (10-24)*

Assuming C2 = C3:

- $$C2 = \frac{1{,}000{,}000}{2\pi F300}$$ *eq. (10-25)*

— where:

C1 and C2 are in microfarads (μF).
F is in hertz (Hz).
R1 is in ohms (Ω).

For best results a multiple CDA integrated circuit such as the LM-3900 should be used for this application. Such a circuit would allow the drift to be controlled because the two halves would both drift at the same rate (they share a common thermal environment).

Using Bipolar DC Power Supplies

The current difference amplifier is designed primarily for single-polarity power supply circuits. In most cases, the CDA will operate with a V+ DC power supply in which one side is grounded. We can, however, operate the CDA in a circuit with a bipolar DC power supply using a circuit such as Figure 10-11. The reference resistor, R_{ref}, is connected from the noninverting input to ground. The V- and V+ power supplies are each ground-referenced and of equal potential. Thus, the 5- to 100-µA bias current is found from $(V+)/R_{ref}$.

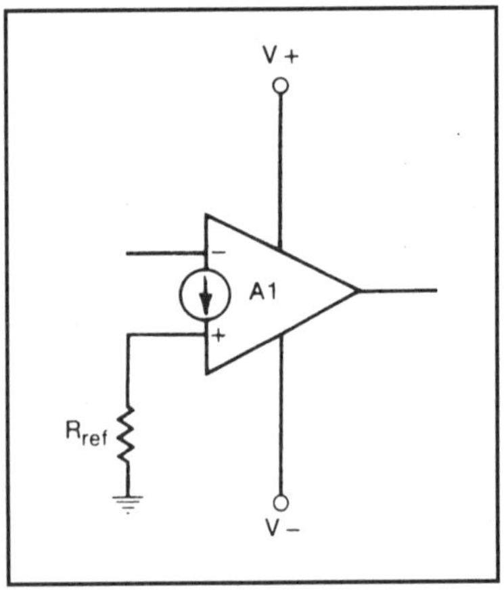

Figure 10-11. Use of a ground-referenced noninverting input.

Chapter 11

Audio Integrated Circuits

The audio spectrum includes those frequencies within the range of human hearing, or about 20 to 20,000 Hz. Audio amplifier circuits range from preamplifiers (which are typically voltage amplifiers dealing with small signals) to high power amplifiers (which drive loudspeakers). In this chapter we will take a look at basic audio amplifiers as implemented in integrated circuit form.

Audio Preamplifiers

The purpose of a preamplifier is to boost the tiny voltage output from sources such as microphones, tape heads, phonograph cartridges and so forth. In audio amplifiers the preamplifier might have a gain of as little as 2 or 3, and as high as 5000. The gain required of the preamplifier is a function of the input signal required by the power amplifier (or other load) and the voltage available from the source. For example, suppose a microphone outputs a voltage of five millivolts (5 mV) at normal speaking levels. Further suppose that the public address power amplifier wants to see a 250-millivolt input signal for maximum output power. We would thus have the V_{in} and V_o potentials of the preamplifier, and could calculate the voltage gain required: $A = V_o/V_{in} = 250$ mV/5 mV = 50.

There are several options in designing preamplifiers for audio use. One popular approach is to use a special purpose integrated circuit (IC) such as the LM-381 device. Others prefer ordinary operational amplifiers, such as the 741 or LM-301 devices. Still others prefer special operational amplifiers designed for audio and stereo applications, such as the LM-1303 device. All of these devices are easily obtained by hobbyists.

DC Power Connections

The IC audio preamplifier may have quite a bit of gain and could thus oscillate under some conditions. It also might easily influence, or be influenced by, other circuits because of signals passed between stages over the DC power lines. Because of these potential problems it is necessary to correctly bypass the DC power lines of the IC preamplifier device. Each power supply line must be bypassed with at least one capacitor, and preferably two. Two of the capacitors are 0.1 µF each, and are used to decouple high frequency signals. The other two capacitors have larger values (typically 4.7 µF) and are used to decouple lower frequency signals. These latter two capacitors are usually tantalum electrolytics.

So why would we want to use two capacitors? Would just one low value capacitor work just as well? The problem is caused by the fact that electrolytic capacitors don't work well at higher frequencies; so we are forced to parallel a lower value capacitor (of a type that is competent at high frequencies) with the higher value electrolytic unit. In many cases, however, we find that the single 0.1 µF capacitor on each DC power supply line is sufficient, so this "overkill" is not needed.

Special Purpose IC Amplifiers

Several companies make IC preamplifiers especially for audio applications. Typical of these are the LM-381 and LM-382 devices (see the National Semiconductor *Linear Data Book* for details). Figure 11-1A shows two circuits based on the LM-381 device. These circuits are intended to show representative applications.

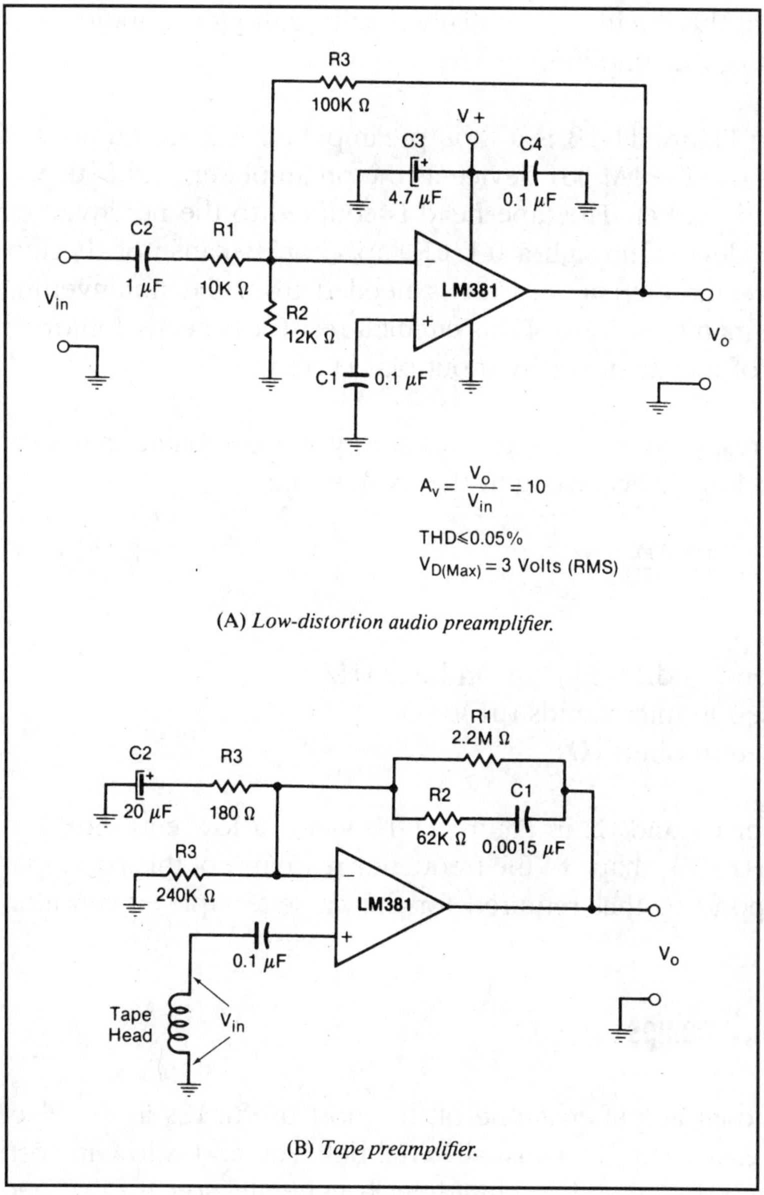

Figure 11-1. A) Low distortion audio preamplifier circuit; B) tape preamplifier circuit.

The LM-381 is similar to a dual-operational amplifier except for the single-polarity DC power supply used. The circuit in Figure 11-1A is a wide band, low-distortion audio preamplifier. With the values shown it will yield a voltage gain of 10 at a total harmonic distortion (THD) of less than 0.05 percent. The frequency response of this circuit is essentially flat throughout the audio spectrum (up to the limit of the amplifier).

The circuit shown in Figure 11-1B is a tape preamplifier. A large number of cassette tape players use the LM-381 device as the preamplifier, and I suspect that was the intended market. The tape head is coupled to the noninverting input of the LM-381 device through a 0.1 μF capacitor. Because of the low input bias of the preamplifier no resistor is needed from the noninverting input to ground. On many versions of this circuit a resistor is needed in order to prevent charging of the capacitor by input bias currents.

The low frequency response of the circuit is set by the combined action of resistor R3 and capacitor C2 according to the expression:

- $$F = \frac{1{,}000{,}000}{2\pi R3C2}$$ *eq. (11-1)*

— where:

F is the low-end -3 dB frequency in hertz (Hz).
C2 is expressed in microfarads (μF).
R3 is expressed in ohms (Ω).

The values shown for R3 and C2 in Figure 11-1B yields a low-end -3dB frequency of about 45 Hz. The shape of the frequency response of this preamplifier is set to correspond to that required for the cassette tape equalization curve.

Operational Amplifier Preamps

The classical operational amplifier is one of the most useful ICs in the electronics catalog. It is easy to apply, is usually well behaved and is low in cost. Designing circuits with the operational amplifier is generally so easy that one guy was tempted to correctly claim that it makes "... the contriving of contrivances a game for all." Circuits that only very advanced hobbyists could tackle are now open to even the newcomer.

Figure 11-2 shows the basic operational amplifier configurations used in audio circuits; Figure 11-2A is the inverting follower and Figure 11-2B is the nonin-

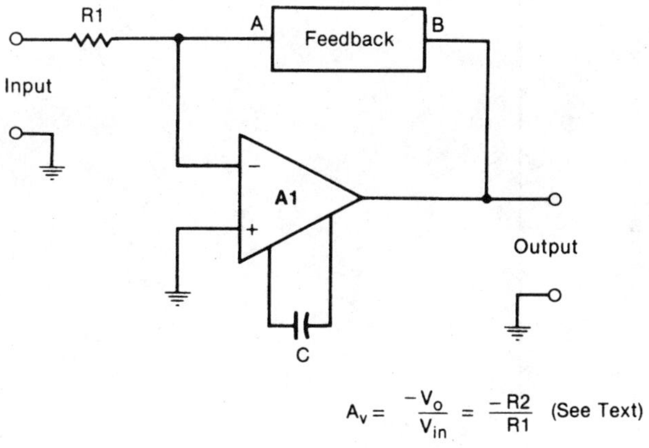

(A) *Inverting feedback amplifier.*

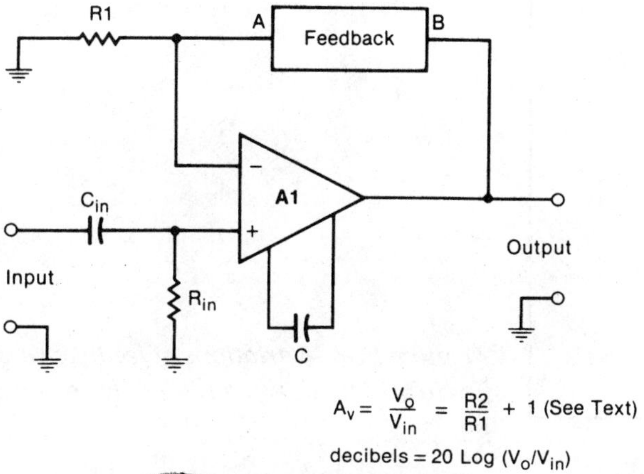

(B) *Noninverting feedback amplifier.*

Figure 11-2. Basic audio configurations for op-amps: A) inverting input feedback amplifier; B) noninverting input feedback amplifier.

verting follower. In the inverting circuit, the output signal will be 180 degrees out of phase with the input signal (i.e., reversed). In the noninverting signal, on the other hand, the output signal is in-phase with the input signal. In both cases, the gain is set by the combined action of the input resistor (R1) and a feedback network. The expressions for the special cases where the feedback network is a single resistor (see Figure 11-3A) are shown with each circuit.

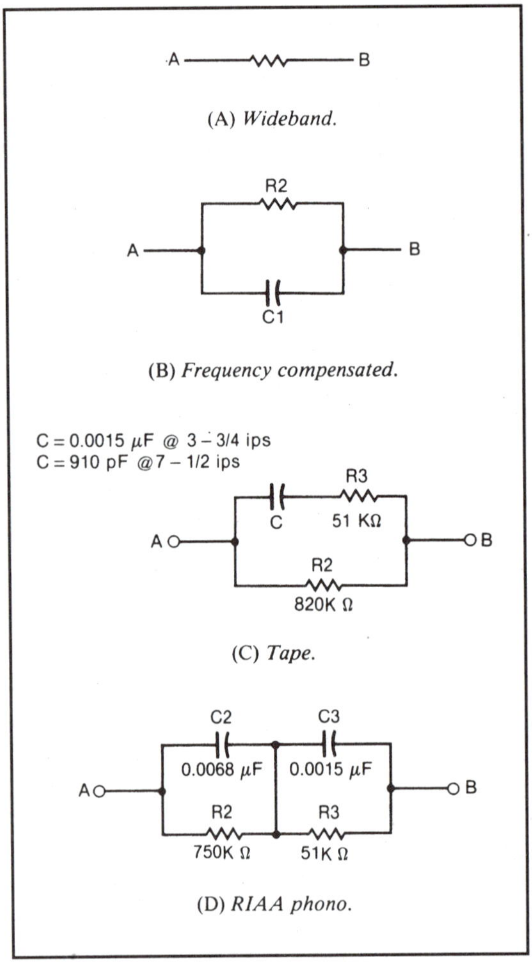

Figure 11-3. Variations of feedback networks used in circuits of Figure 11-2.

Several popular variations of the feedback network that find application in audio preamplifiers are shown in Figure 11-3. The version in Figure 11-3A is for a wide band amplifier that has a frequency response with no tailoring except by the natural bandwidth of the amplifier. We can calculate the approximate available bandwidth from the gain-bandwidth product (F_t) of the device, which is the frequency at which gain drops to unity (1). The expression is:

- $$F_t = GAIN \times BANDWIDTH \qquad \text{eq. (11-2)}$$

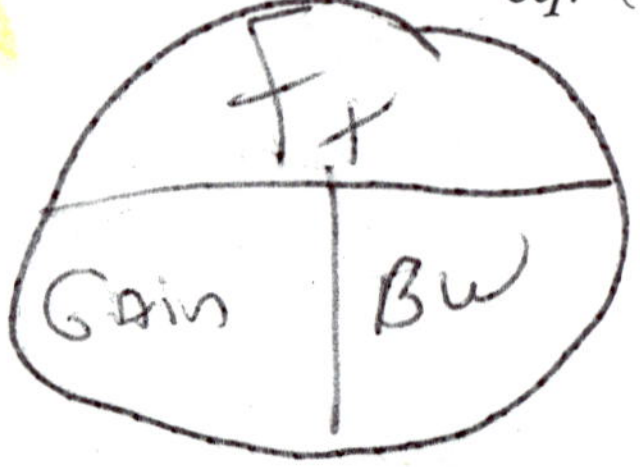

BW = upper Audio freq

One use of this expression is to determine which op amp is needed. For example, suppose we need a frequency response of 20,000 Hz in an amplifier with a gain of 150. The required F_t is 150 x 20,000 = 3 MHz.

Alternatively, we could also rearrange the expression to calculate the maximum frequency response of any given circuit (if we know F_t). For example, suppose we have a 3 MHz op amp in a circuit with a gain of 100. The maximum frequency response of the amplifier will be F_t/Gain = 3 MHz/100 = 30 kHz.

The feedback network shown in Figure 11-3B produces a flat gain at low frequencies equal to that of the resistor alone. At frequencies above a certain point, however, the gain of the amplifier will roll off at -6 dB per octave. The "break point" between the low frequency gain and the rolled-off segment (i.e., the high-end -3 dB point) is:

$$F = \frac{1{,}000{,}000}{2\pi R2C1} \qquad \text{eq. (11-3)}$$

— where:

F is in hertz (Hz).
R2 is in ohms (Ω).
C1 is in microfarads (μF).

The remaining feedback networks are used in special preamplifiers. That shown in Figure 11-3C is used for NAB-compensated tape preamplifiers, while that of Figure 11-3D is used for RIAA-compensated phonograph preamplifiers.

Operational amplifiers are normally used in dual-polarity power supply circuits. In cases where the device must be used in single-supply circuits we can use a configuration such as Figure 11-4. In this type of circuit the V- power supply terminal is grounded, and the V+ terminal is connected to the single power supply. The noninverting input is biased to a point midway between V+ and ground by a voltage divider consisting of R3 and R4. The value of R3/R4 can be any resistance between 2K and 100K, with the value shown (3.3K) being common.

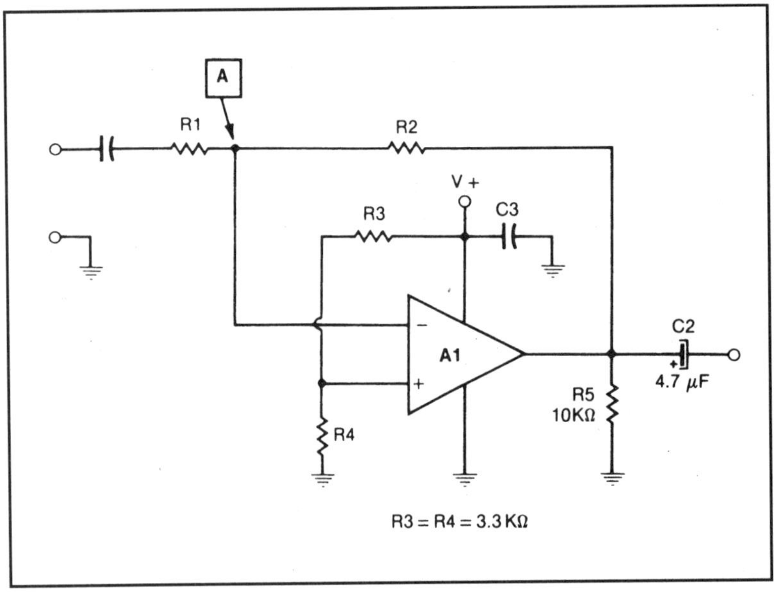

Figure 11-4. Single DC supply audio preamplifier.

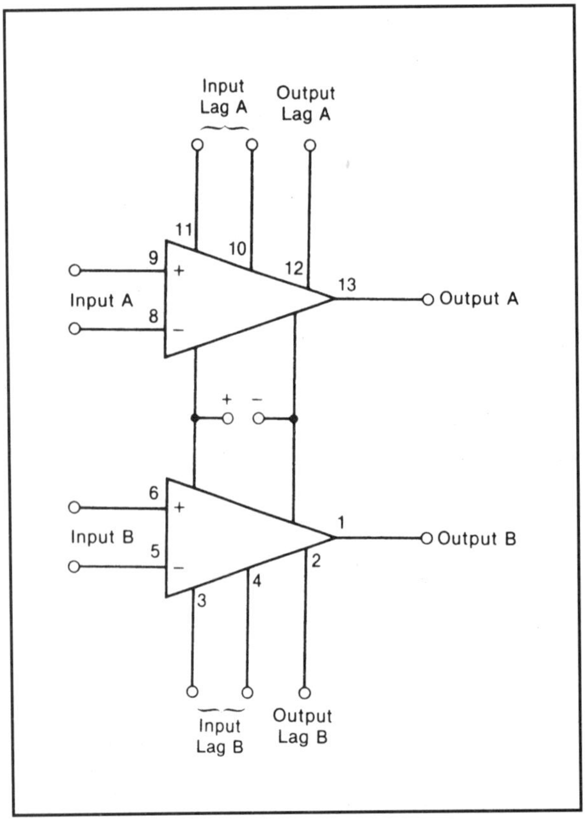

Figure 11-5. LM-1303 stereo audio preamplifier chip.

A problem with this circuit is that the inverting input and the output terminals are also biased to a high DC potential. In order to prevent these voltages from affecting other circuits we need to capacitor couple this circuit. Capacitor C1 couples the input signal to the amplifier while preventing the DC bias at point "A" from affecting the input source. Similarly, capacitor C2 will pass the audio output signal while blocking the DC offset bias.

Figure 11-5 shows an example of a stereo audio preamplifier based on the op amp. It was once called the MC-1303 when Motorola was the only source of the chip; but others now make it so it is now available from many sources under the type number LM-1303. The chip contains two operational amplifiers that are completely independent except for the V- and V+ DC power supply connections.

Tone Control Circuits

Tone controls allow the user to custom tailor the frequency response of an audio preamplifier. The bass control emphasizes the low frequencies, while the treble control emphasizes the high frequencies. In some circuits a single control serves both ends of the spectrum, while in others separate bass and treble controls are provided. Figure 11-6 shows several varieties of tone control circuit.

The circuit in Figure 11-6A is a simple treble roll-off tone control found in low cost equipment. This is the simplest and least desirable variety, and consists of two components series-connected across the audio line. These components form a treble roll-off circuit that mimics the effect of "bass boost" (poorly) by rolling off the high frequencies. Unfortunately, while simple, this type of tone control reduces the signal amplitude.

Another variety of tone control is shown in Figure 11-6B. In this version a pair of frequency-sensitive RC or RLC networks is placed in the negative feedback circuit of the amplifier. These circuits selectively amplify different bands of frequencies. An example of the RLC version is shown in the inset in Figure 11-6B. This circuit is popular in graphic equalizers.

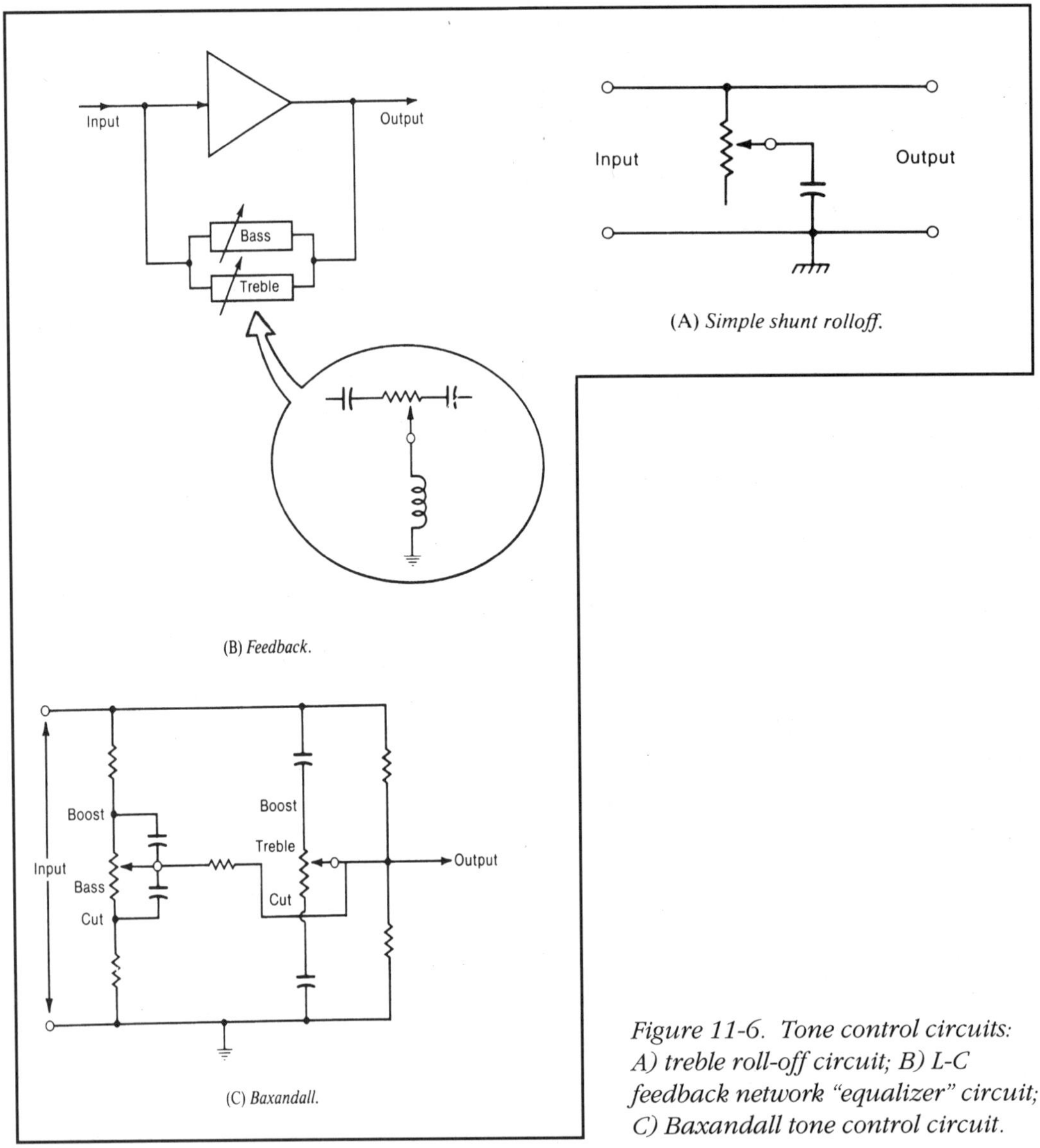

Figure 11-6. Tone control circuits: A) treble roll-off circuit; B) L-C feedback network "equalizer" circuit; C) Baxandall tone control circuit.

The Baxandall tone control circuit shown in Figure 11-6C is used in almost all decent audio amplifiers. The Baxandall circuit consists of two cascade frequency selective RC networks, one each for bass and treble ranges. Each control will boost its design frequencies at one end of the control's range, and will cut it at the other end of the range.

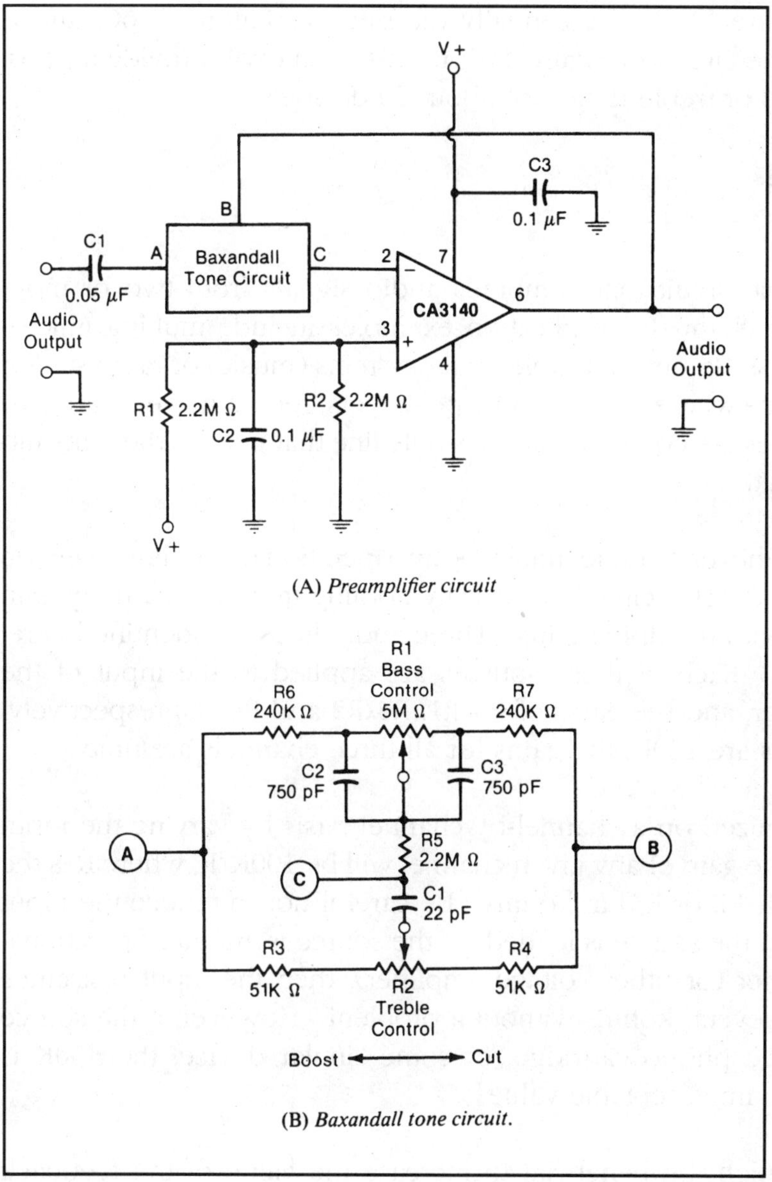

Figure 11-7. Tone controlled preamplifier stage: A) preamplifier circuit; B) Baxandall tone network.

Figure 11-7A shows the circuit for a tone control preamplifier stage based on the Baxandall circuit. The basic circuit is a CA-3140 operational amplifier connected in a basic inverting follower configuration, with a single power supply. Resistors R1 and R2 bias the noninverting input to (V+)/2. These resistors are higher in value than in certain other op amps because of the very large input impedance of the BiMOS CA-3140 op amp. The three-terminal "Baxandall tone circuit" of Figure 11-7B is essentially the same as Figure 11-6C, and is used as shown in the circuit of Figure 11-7A. This circuit will provide a boost or cut of either bass or treble ranges of about 20 decibels.

Audio Mixer Circuits

An audio mixer is a circuit that combines audio signals from two or more inputs into a single channel. Application examples include multiple microphone public address systems, multiple guitar systems (music) or radio station audio console service where inputs from tape players, record players, and two or more microphones are combined into a single line that goes to the transmitter modulation input.

Audio Mixer I. The first audio mixer is an operational amplifier version shown in Figure 11-8. This circuit is basically nothing more than a unity gain inverting follower with multiple inputs. Three audio lines are identified here: AF1, AF2 and AF3. Each of these sources are applied to the input of the operational amplifier, and see gains of R4/R1, R4/R2 and R4/R3, respectively. Because all resistors are 100K, the gains for all three channels are unity.

Gain can be customized on a channel-by-channel basis by varying the input resistance value. The gain of any given channel will be 100K/R, where R is the input resistance (R1, R2 or R3) in kohms. Be careful not to reduce the input resistance so far that the source is loaded. If the source is another operational amplifier preamplifier (or other voltage amplifier), then the input resistance can be reduced to several kohms without a problem. However, if the source is a high impedance phono cartridge, or some similar device, then 50K is probably the minimum acceptable value.

In some cases, it might be beneficial to increase the value of the feedback resistance to 1 megohm or so, in order to make the corresponding input resistances higher for any given gain. Remember, the input impedance seen by any single channel is the value of the input resistor.

A master gain control is provided by making the feedback resistor variable from 0 to 100K. If no gain control is needed, then make this resistor fixed. An

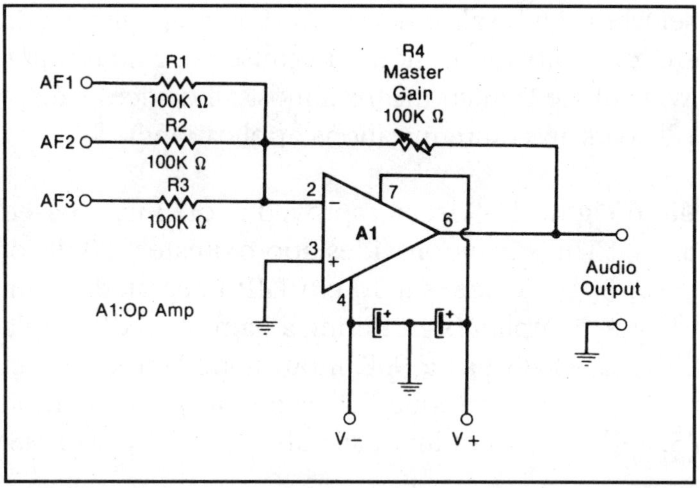

Figure 11-8. Three-input audio mixer with master gain control.

audio taper potentiometer is used for most applications. If the application calls for a one-time "set-and-forget" gain control adjustment (as might be true in radio station applications), then make R4 a trimmer potentiometer; otherwise it should be a panel type with a shaft appropriate for a knob (quarter-inch half or full round).

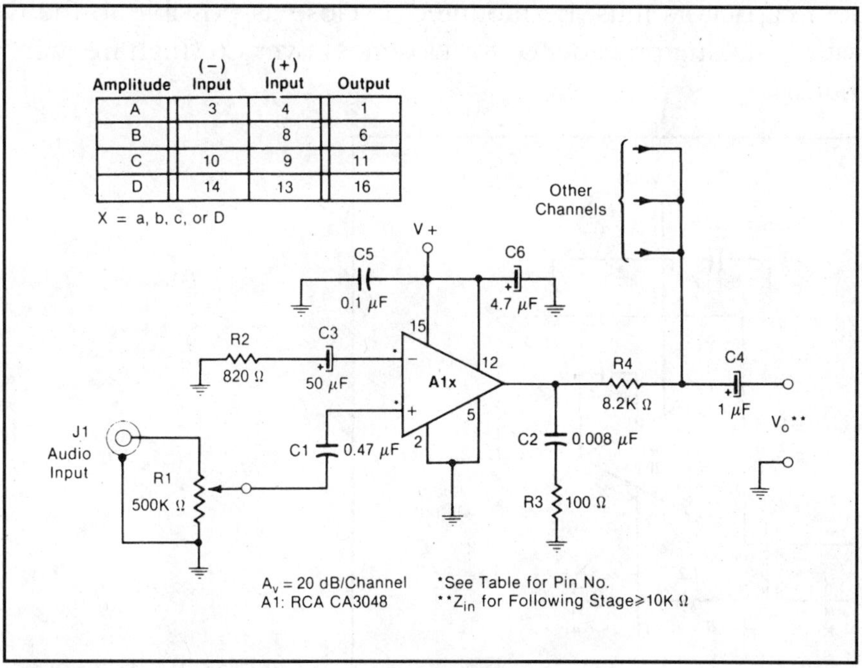

Figure 11-9. ca-3048 four-channel mixer circuit.

The operational amplifier selected can be almost any good op amp with a gain bandwidth product sufficient for audio applications. Because the gain is unity, any GBP over 20 J = kHz will suffice (which means almost all devices except the 741 family — which will work in communications applications).

Audio Mixer II. The circuit in Figure 11-9 is an improved audio mixer based on the CA-3048 amplifier array. This circuit provides approximately 20 dB of gain at each channel. The CA-3048 device is a 16-pin DIP integrated circuit that contains four independent AC amplifiers. Offering a gain of 53 dB with a GBP of 300 kHz (typical), the CA-3048 has a 90K input impedance, and an output impedance less than 1K. It will produce a maximum low-distortion output signal of 2 volts RMS, and can accept input signals up to 0.5 volts RMS.

Each DC power supply can be up to +16 volts DC. There are two V+ and two ground connections. These multiple connections are used to reduce the internal coupling between amplifiers. The two V+ terminals are tied together externally, and the two ground terminals are tied together also. The V+ terminals are bypassed with a dual capacitor. C5 is a 0.1 µF unit, and is used for high frequencies, while C6 is a 4.7 µF tantalum electrolytic for low frequencies. Both capacitors must be mounted as close as possible to the CA-3048 body, with C5 taking precedence for closeness over C6 (high frequencies are more critical).

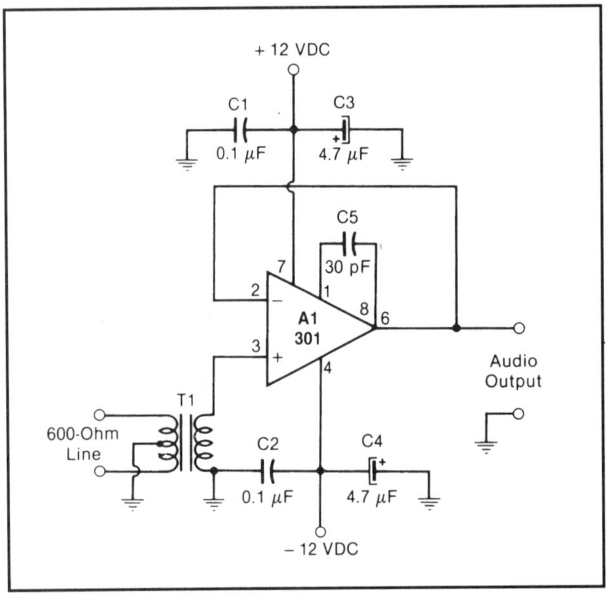

Figure 11-10. Transformer coupled line amplifier.

An RC network from the amplifier output to ground (R3/C2) is used to stabilize the amplifier, thus preventing oscillation. Like the power supply bypass and decoupling capacitors, these components need to be mounted as close as possible to the body of the amplifier.

Only one channel is shown in detail in Figure 11-9 because of space limitations. However, each of the other three channels are identical, and are joined with the circuitry shown at the output capacitor (C4) as shown. Each of the four channels has its own level control (R1), which also provides a high input impedance for the mixer.

600 Ohm Audio Circuits

Professional audio and broadcast applications generally use a 600 ohm balanced line between devices in the system. For example, a remote preamplifier will have a 600 ohm balanced output and will connect to the next stage through a three-wire line. Such a system uses two "hot" lines and a ground as the stage-to-stage interconnection. An amplifier with a 600 ohm balanced output is called a line driver amplifier, while an amplifier with a 600 ohm balanced input is a line receiver. Some amplifiers, of course, are both line drivers and line receivers.

Figure 11-10 shows a line receiver amplifier based on the LM-301 operational amplifier used in the unity gain, noninverting configuration. The input circuit is a line transformer. If the turns ratio of the transformer T1 is 1:1, then the overall gain of the circuit is unity. However, if the transformer has a turns ratio other than 1:1, then the "gain" is essentially the turns ratio of the transformer. For example, suppose a transformer is selected with a 600 ohm balanced input winding, and a 10,000 ohm secondary winding. Such a transformer will have a sec/pri impedance ratio of 10,000/600, or about 17:1, so the turns ratio is on the order of the square root of 17, or about 4.1. Thus, the voltage actually applied to the noninverting input will be 4.1 times higher than the input signal voltage from the 600 ohm line.

Like most other op amp circuits, Figure 11-10 requires two DC power supplies (V- and V+). Typically, these power supplies can be anywhere from ± 6 to ± 15

volts. As in other circuits, each DC power supply is decoupled with a pair of capacitors, a 0.1 µF for higher frequencies and a 4.7 µF unit for lower frequencies.

The output of this circuit is an ordinary single-ended voltage amplifier (as in other op amp circuits), so will typically have a very low impedance. Some designers use a transformer-coupled output circuit. It is possible to get away with a 600 ohm 1:1 transformer if the natural output impedance of the op amp is on the order of 50 ohms or so. The general rule of thumb is that the primary impedance of any transformer selected should be 10X the natural output impedance of the device, for best voltage transformation.

Another way to make a 600-ohm line input amplifier is to use the simple DC differential circuit in Chapter 6, and make sure the input resistors are 300 ohms each (note: 330 ohms is the closest easily-obtained standard value of a simple resistor, and will work fine in most cases).

Figure 11-11 shows a line driver amplifier based on a pair of operational amplifiers (the DC power supply connections are deleted, but all pin-outs are as in Figure 11-10). The output circuitry is balanced because it is made from two single-ended op amps driven out of phase with each other. The low output impedance of the operational amplifier, plus the 270 ohm series resistance, makes the balanced output impedance a total of approximately 600 ohms.

The circuit of Figure 11-11 is a good example of the clever use of one of the properties of the ideal op amp. Recall that one of the ideal properties is that inputs "stick together." In other words, applying a voltage to one input causes the same voltage to appear at the other input. In this case, for example, the AF input signal voltage applied to the noninverting input of amplifier A1 also appears on the inverting input of that same amplifier. Thus, we will see V_{in} both on the noninverting input and at point "A." We can therefore use point "A" to feed the other half of the balanced circuit, amplifier A2. Because A1 is a noninverting gain-of-two circuits, and A2 is a gain-of-two inverting circuits, the two sides are out of phase with each other — which is the condition required of the two "balanced" output lines.

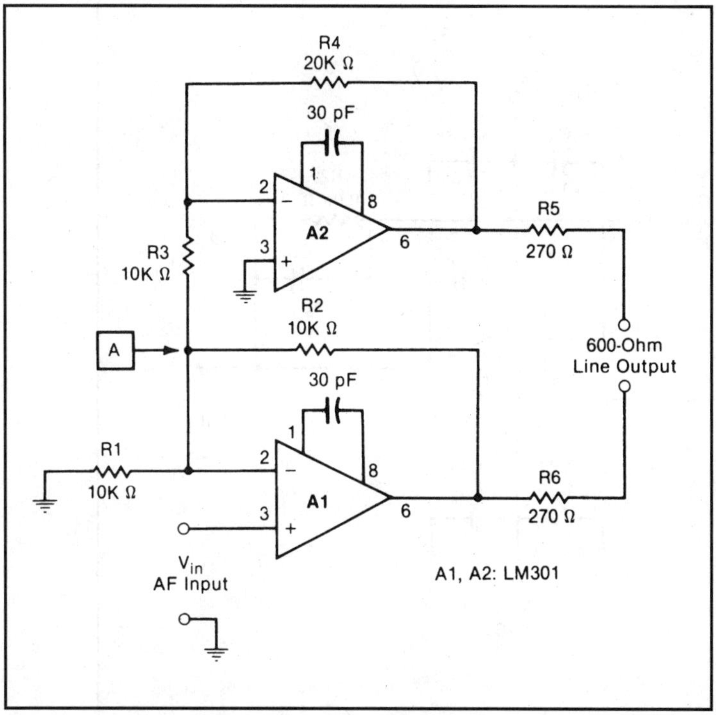

Figure 11-11. Transformerless line driver amplifier.

Some Practical Preamplifier Circuits

A preamplifier circuit is an audio amplifier that gives some initial amplification to the signal before passing it to another circuit for additional amplification or other processing. For example, a microphone has a low-level output (several millivolts to 0.2 volts). A preamplifier typically boosts the microphone signal to 100 to 1000 millivolts before it is applied to the input of a power amplifier (for the loudspeaker) or the input of a transmitter modulator (depending upon use).

Microphone Preamplifier. Figure 11-12 shows a simple gain-of-100 microphone preamplifier for communications and non-Hi Fi public address uses. This circuit uses an LM-301 operational amplifier in the noninverting follower gain configuration. With the values of feedback and "input" resistors (R1 and R2) shown, the gain is 101.

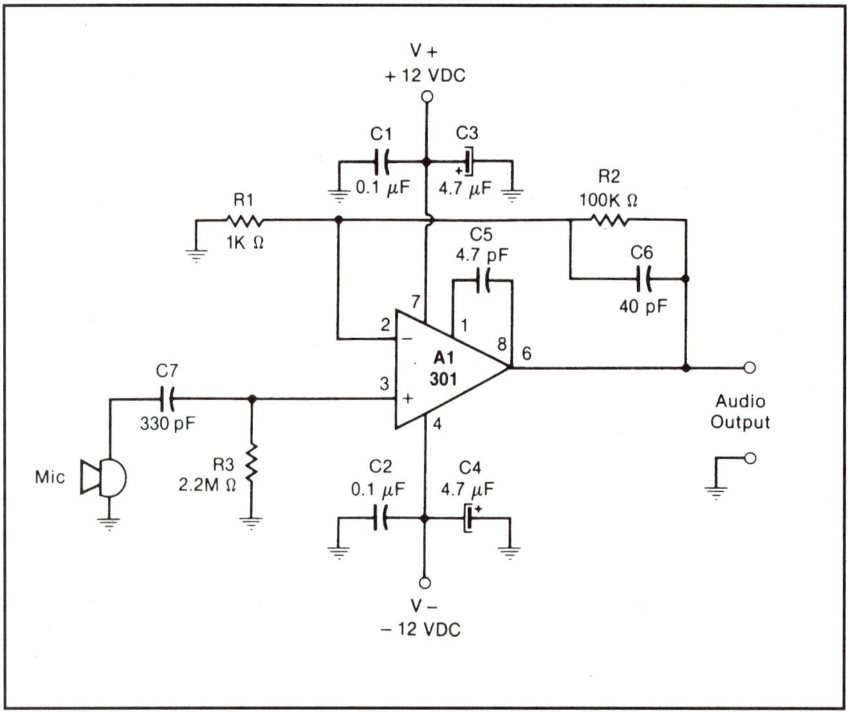

Figure 11-12. Microphone preamplifier.

The input circuit of this preamplifier is capacitor-coupled to the microphone. In order to keep the input bias currents of the op amp from charging the capacitor (and thereby latching up the op amp), we use a 2.2 megohm resistor (R3) from the noninverting input to ground. This circuit is relatively general, but can be modified to be less complex if we use a dynamic microphone only. Those forms of mike use either a high- or low-impedance coil (like a loudspeaker, but reversed in function) that is permanently connected into the circuit. In that case, delete R3 and C7, and connect the microphone between ground and pin no. 3. If the microphone is to be disconnected from time to time, however, we must keep R3 in the circuit to prevent the op amp output from saturating at or near V+ when the noninverting input "goes open."

The frequency response of this circuit is tailored by C5 and also by the capacitor shunting the feedback resistor (C6). With the values of capacitance shown, the upper -3 dB point in the response curve is a little over 3,000 Hertz, and a falls off at a rate of approximately -6 dB/octave above that frequency.

Audio Compression Amplifier

A compression amplifier is one that reduces its gain on input signal peaks, and increases the gain in the signal valleys. These circuits are used by electronic music fans, and by broadcasters, to raise the average power in the signal without appreciable distortion. Figure 11-13 shows a typical compression amplifier circuit.

The amplifier (A1) is any good audio operational amplifier, such as the LM-301 (power supply and compensation not shown, but see earlier circuits in this chapter). The gain of the circuit is set by the input resistor (R1), and a feedback resistance that consists of the parallel combination of R2 and the resistance of the optocoupler (OT1) output element. The OT1 resistance is set by the intensity of the LED brightness; which is, in turn, set by the signal amplitude produced by A2. Because the A2 output signal is proportional to the A1 output signal, overall gain "reduces" itself (i.e., compresses). Optoisolator OT1 can be any resistance output device, such as the Clairex, or a modern type (e.g., H11) that uses a JFET for the resistance element.

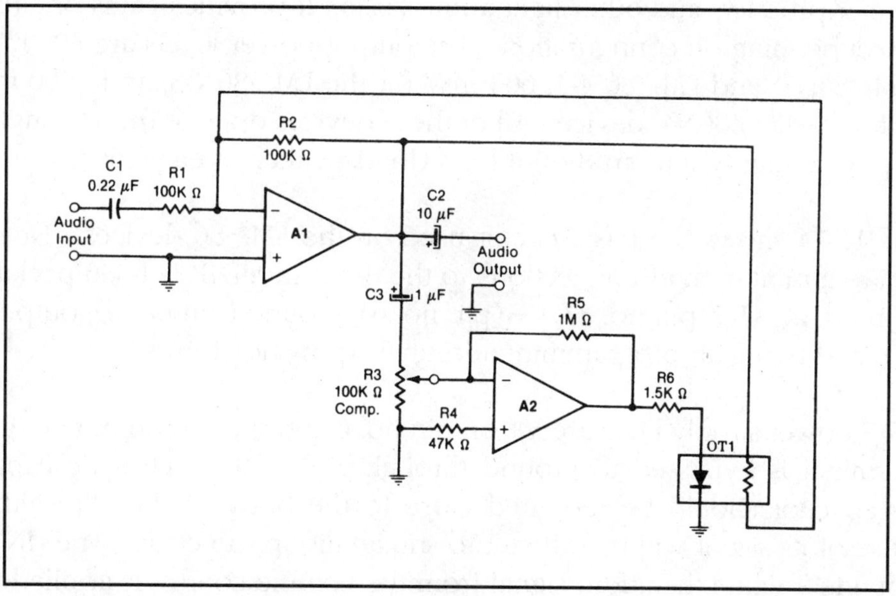

Figure 11-13. Audio compression amplifier.

IC Audio Power Amplifiers

The circuits discussed thus far in this chapter are basically audio voltage amplifiers. They produce an increase in signal voltage as their function. Even unity gain voltage amplifiers produce an increase in power level because the output impedance is so much lower than the input impedance. Thus is it for all of the circuits in this chapter. However, the maximum power level is only on the order to 20 to 100 milliwatts, which is hardly useful. An operational amplifier (or other linear IC audio circuit) can drive an audio load, such as one of those tiny little earphones used on the smallest transistor radios. When more power is needed, we turn to other IC devices. The power levels of these chips range from 200 mW to 200 watts; but for our purposes, we will look at the low-power LM-386 to the medium-power LM-2002. Consulting the data books from semiconductor manufacturers will show a large variety of other devices at all power levels.

The LM-386 Audio Preamplifier/Power Amplifier

The LM-386 is an entire audio system designed as a single IC for portable consumer products and other light audio tasks. It provides gains of either 20 or 200 depending on configuration. The output power levels are 300 mW for the LM-386N-1 and LM-386M-1, 600 mW for the LM-386N-3, and 1000 mW (1 watt) for the LM-386N-4 device. All of these devices operate from a single +12 VDC power supply into an 8-ohm load (loudspeaker or earphones).

Figure 11-14 shows the basic circuit used on the LM-386 devices. Note that there are a minimum of connections to the 8-pin miniDIP LM-386 package: -IN (pin no. 2), +IN (pin no. 3), V+ (pin no. 6), ground (pin no. 4), output (pin no. 5), and two gain programming terminals (pins no. 1 and 8).

The DC power supply lines are V+ on pin no. 6 and ground on pin no. 4. The V+ terminal is bypassed to ground through a 470 µF electrolytic capacitor. This capacitor should be mounted close to the body of the clip; although because of its size it will be difficult to snuggle it up too close. The -IN input is grounded, while the audio signal from the volume control is applied to the +IN terminal (pin no. 3).

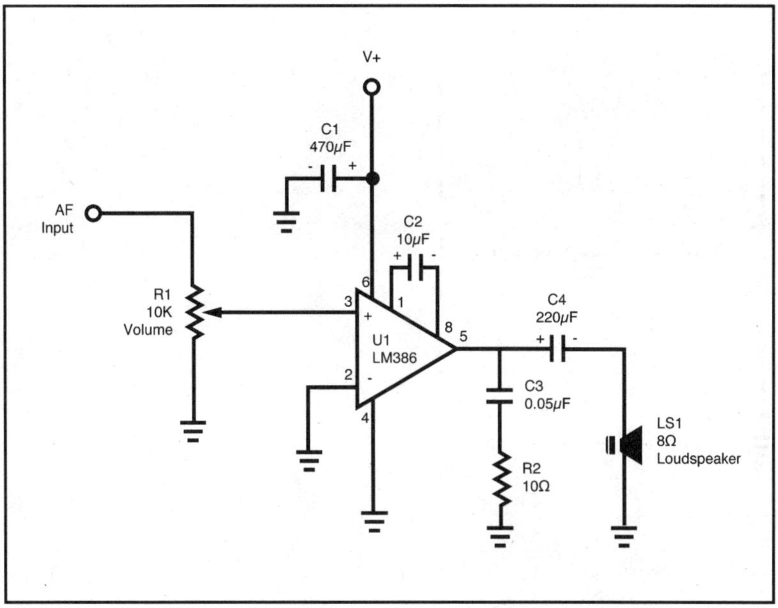

Figure 11-14. LM-386 audio power amplifier can be used for small receiver projects.

Output from the LM-386 is passed through a DC blocking capacitor (C4) to the 8 ohm loudspeaker. When the LM-386N-4 (1000 mW) unit is used, then a snubbing network (R2/C3) is used to prevent self-oscillation.

The gain of the LM-386 is dependent on whether or not capacitor C2 is used. When this 10 µF capacitor is in the circuit, the gain of the LM-386 is X200; but if it is deleted, then the gain is X20.

The LM-2002 IC Audio Power Amplifier

The LM-2002 device looks like a plastic power transistor with five terminals rather than three. It will produce up to 8 watts of audio into a 4-ohm loudspeaker load. Note in Figure 11-15A that the circuit is minimal because the device has only -IN, +IN, V+, GND and OUT terminals. The terminal number is shown in Figure 11-15B, and is given below:

Pin	Function
1	-IN Input
2	+IN Input
3	Ground

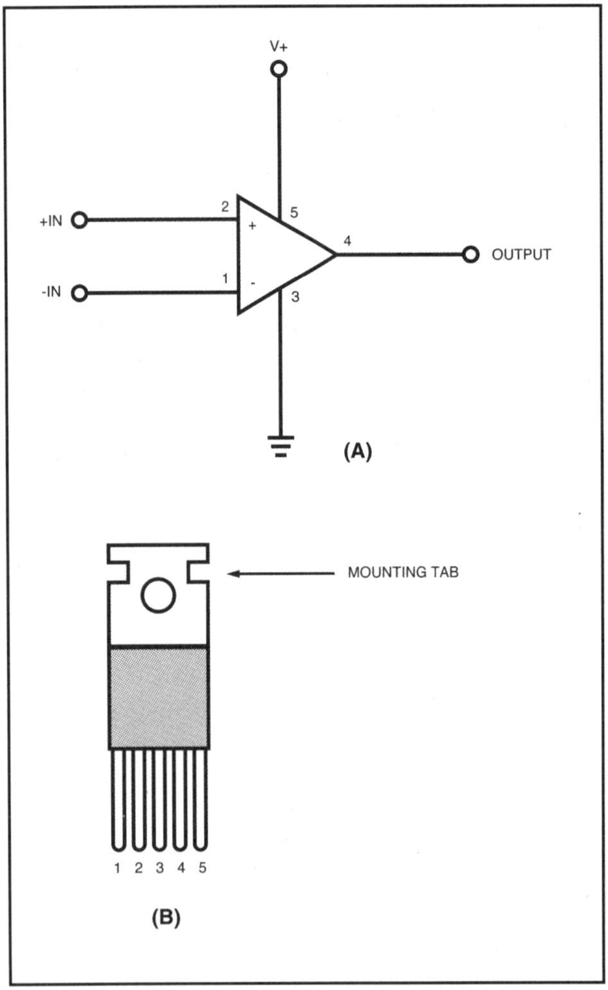

Figure 11-15. LM-2002 8-watt audio amplifier device: A) circuit symbols and pin-outs; B) package.

4	Output
5	V+ DC Power

Notice that the circuit is vaguely similar to the LM-386, but the inverting input (-IN) can be programmed like an op amp for the overall gain of the circuit. See Figure 11-16 for the circuit connections.

When using the LM-2002, be aware that the LM-2002 device is a power amplifier that delivers up to 8 watts. While the LM-386 can be used as-is, the LM-2002 must be mounted to a heat sink. The LM-2002 manufacturer recommends the Staver V-5 heat sink, but similar devices are also acceptable. The mounting tab is also the heat transfer surface. Use a machine screw through the mounting hole to secure the LM-2002 to the heat sink. Better hear transfer will occur when a thin film of silicone heat transfer grease is used between the mounting tab and the heat sink.

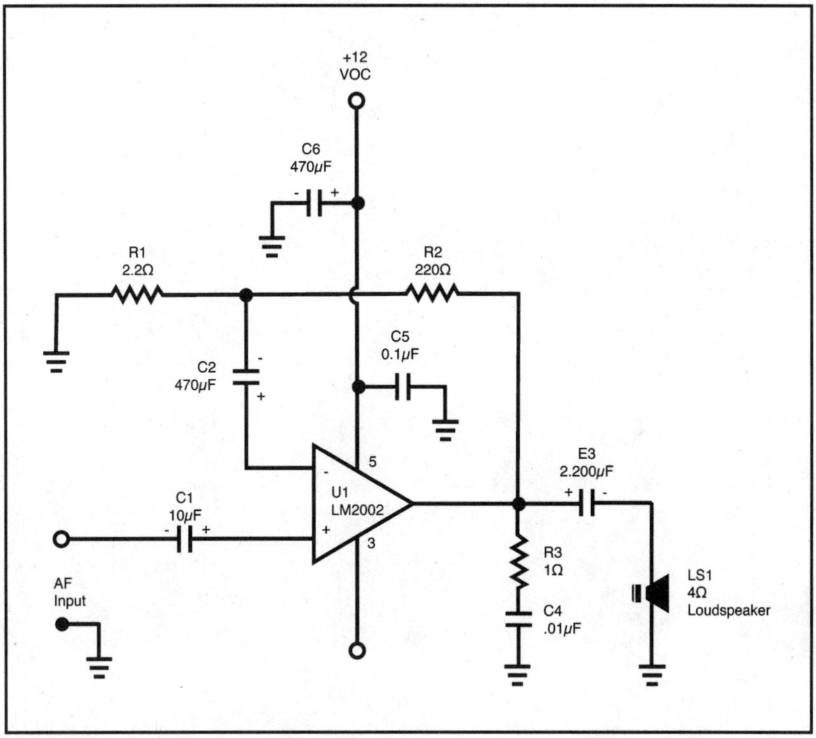

Figure 11-16. LM-2002 circuit connections.

Chapter 12

Wideband and Video Linear IC Amplifiers

A wideband amplifier is one that has a frequency response over a wide band (how's that for a bit of circular logic?). Seriously, most of the amplifiers that we have discussed thus far are relatively narrow-band devices. Indeed, when an operational amplifier has a wide bandwidth, it tends to oscillate unless some external components are added. A video amplifier is a special case of a wideband amplifier used for television-type signals. For purposes of this chapter, a wideband amplifier is any amplifier, including video amplifiers, that has a frequency response bandwidth of 500 kHz or more.

Most wideband amplifiers have 50-ohm input and output impedances, while video amplifiers typically have 75-ohm input and output impedances. These impedance values are considered standard for RF and TV circuits, respectively.

General Wideband Preamplifier

Figure 12-1 shows two general purpose preamplifiers based on the CA-3600E device. This IC is a complementary COS/MOS transistor array. A single-stage design is shown in Figure 12-1A, while a multistage design is in Figure 12-1B.

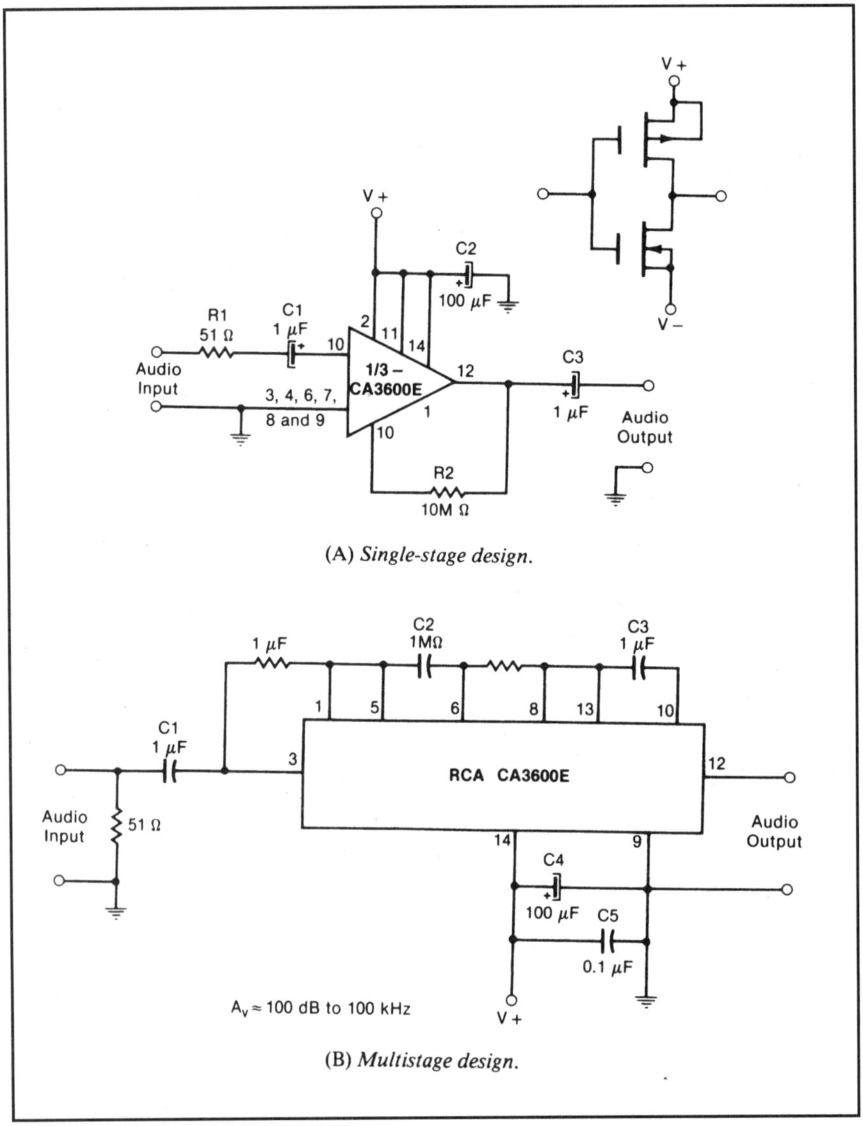

Figure 12-1. General purpose wideband amplifier based on CA-3600E: A) single-stage design; B) multistage design.

The internal transistor array equivalent circuit for one transistor pair is shown in the inset to Figure 12-1A. The single-stage design is capable of up to 30 dB of gain at a V+ of 15 volts DC, and slightly more at lower potentials (but only at a sacrifice of the 1 MHz -3 dB point).

The multistage design shown in Figure 12-1B is capable of voltage gains to 100 dB at frequencies up to 1 MHz (assuming a 10 volt DC supply — gain drops to 80 dB for +15 volt supplies). This gain and frequency response is quite useful in audio and other applications, but must be approached with caution when actually built. Be sure to keep the input and output sides of the amplifier separated as much as possible. Also make sure that the power supply decoupling capacitors are mounted as close as possible to the body of the IC.

MAR-1 Wideband Preamplifier

Building a really wideband preamplifier that works over, say, near-DC to 1,000 MHz, was always a real torturous chore, until recently. The MAR-x series of wideband chips from Mini-Circuits makes the job really easy. These devices are low cost, and offer the RF and wideband amplifier builder a real advantage: they inherently have 50-ohm input and output impedances (needed for RF systems).

The amplifier of Figure 12-2 is "minimalist," but works well. The MAR-1 device, a member of the MAR-x family, offers >13 dB gain from near-DC to 1000 MHz, so will work throughout the HF shortwave, VHF and UHF bands.

The MAR-1 is simplicity itself, for there are only four terminals: input (pin no. 1), output (pin no. 3) and two grounds (pins 2 and 4). The inset to Figure 12-2 shows the MAR-1 package, which resembles a small-signal UHF/microwave transistor. Pin no. 1 is identified by two things: its lead is beveled (the other leads are rounded on the ends) and there is a brown color dot at pin no. 1 (other MAR-x devices use different color dots).

The V+ DC power is applied to the output terminal (pin no. 3) through a current-limiting resistor (R1) and an optional RF choke (RFC1). The value of the resistor is set by Ohm's law, and by setting the applied voltage (V) to a value between 5 and 7 volts (higher voltages will harm the unit). The value of R1 is:

- $$R1(Ohms) = \frac{(V+)-V}{0.015} \qquad eq.\ (12\text{-}1)$$

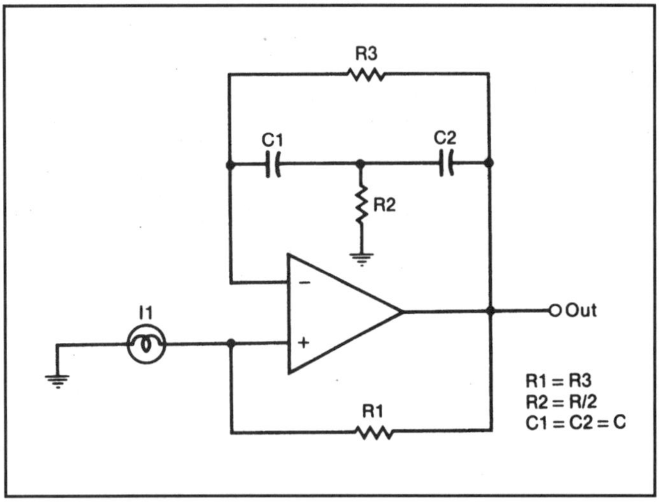

Figure 12-2. MAR-1 near-DC to 1,000 MHz amplifier (inset: MAR-1 package).

(Note: for the MAR-1, V is 5 volts, for others in the series it goes up to 8 volts. Consult the MiniCircuits catalog).

Equation (12-1) assumes a current (I) of 15 mA (0.015 A), which is proper for the MAR-1 (other MAR-x devices may use different optimum currents). If the V+ voltage is +9 VDC, and the applied MAR-1 voltage is +5 VDC, then R1 is 270 ohms.

The input and output capacitors should be 1000 pF (0.001 µF) under 50 MHz (some say 100 MHz), and 100 pF above that frequency. When VHF/UHF operation is contemplated, then C1 and C2 should be "chip capacitors," otherwise disk ceramic capacitors are satisfactory. Capacitor C3 is for power supply decoupling, and should be a disk ceramic model. Use 0.01 µF for HF and low VHF, and 0.001 µF above 50 MHz.

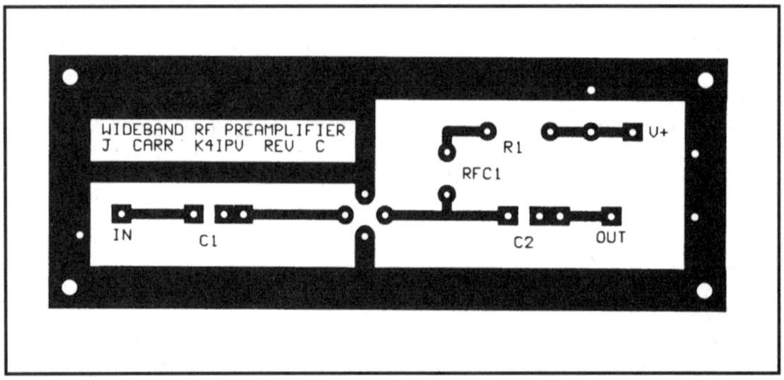

Figure 12-3. Printed circuit pattern for MAR-1 amplifier.

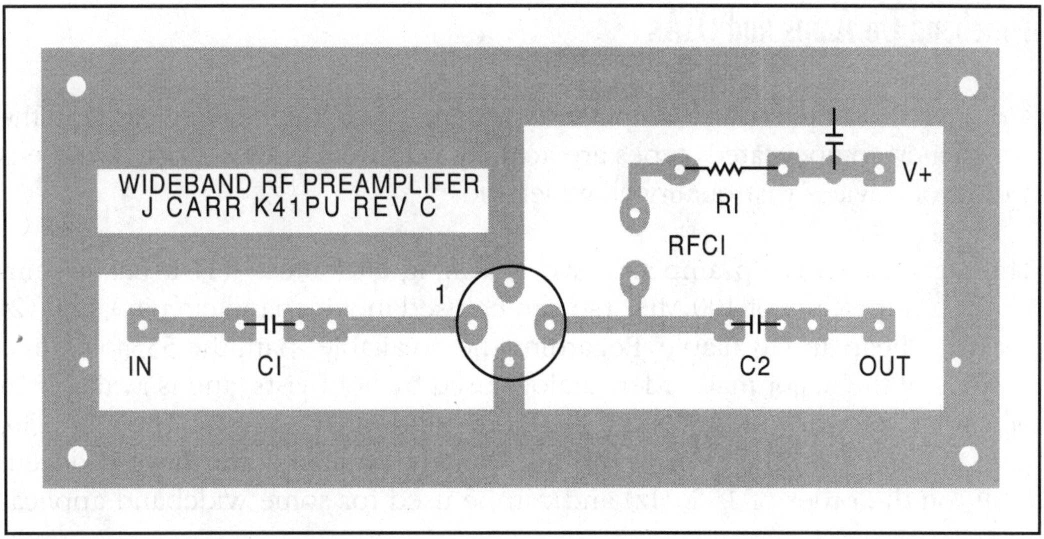

Figure 12-4. Components layout for printed circuit board.

Figure 12-3 shows a printed circuit pattern for this project, while Figure 12-4 shows the components layout. The sites for C1 and C2 are designed to accommodate several different capacitor types. The spacing between the ends of the PC track is set to accommodate chip capacitors. These capacitor are soldered directly to the foil side of the printed circuit board. On one side of each capacitor site there are two holes to accommodate disk ceramic or other high frequency capacitors. The reason for using two holes is that lead spacing is different for different capacitors (even in disk ceramics). As a result, I selected the two most common spacings, and accommodated both of them. Pick the one that matches the capacitors you use (non-chip capacitors are mounted on the non-foil side of the PCB).

The MAR-1 device is mounted in either of two ways. First, you can mount the MAR-1 directly to the foil side of the PCB (note where pin no. 1 is located). This approach is recommended for upper UHF scanner operation, or in the VHF ham bands above 220 MHz. Second, you can bend the leads 90 degrees, and insert them through the holes provided from the component (non-foil) side of the board. This method will work well into the VHF/UHF region, but is not recommended for higher UHF service. When you bend the leads of the MAR-1, bend them away from the color dot side (top) of the package so that you will not lose track of where pin no. 1 is located.

Wideband Op Amps and OTAs

Although most operational amplifier types are low frequency devices, and the "frequency compensated" types are actually very frequency restricted, there is a class of devices that qualify as wideband.

The Signetics 5534 op amp has two charming attributes: (1) it has a gain-bandwidth product of 100 MHz (so can be used in wideband circuits), and (2) it's both cheap and available. Regarding the "available" part, the 5534 is found in some of the major mail-order catalogs used by hobbyists, and is available in blister pack form from Jameco ("Jim-Paks") and other sources. The CA-3130, CA-3140 and CA-3160 devices are also widely available, and have a decent GBP (on the order of 15 MHz) and can be used for some wideband applications.

Burr-Brown offers a series of operational amplifiers and an operational transconductance amplifier that have some really exciting GBP specs. The B-B OPA-622AP has a GBP of 150 MHz, and the OPA-622AU is 200 MHz; both in a 14-pin DIP package. The OPA-627 and OPA-637 are available in metal can and 8-pin miniDIP packages, with 16 MHz and 80 MHz GBP, respectively. A pair of 200 MHz GBP devices are the OPA-678 in 16-pin DIP, and OPA-620 in 8-pin miniDIP. The OPA-678 is notable because it has two sets of differential inputs ("A" and "B") with a TTL A/B switching line to determine which input is used at any one time. The OPA-623 is an 8-pin miniDIP device with a GBP of 350 MHz, while its cousin, the OPA-621 device is the same sort of thing with a 500 MHz GBP. An operational transconductance amplifier, the OPA-2662, offers OTA performance with a GBP of 350 MHz. Another interesting product is the BUF-634 buffer amplifier. It comes in two package styles. One is the 8-pin miniDIP, while the other is a tab mounted "power transistor" case that looks like a TO-220 with five leads (similar to the LM-2002 case shown in Chapter 11). The BUF-634 has a wideband mode up to 180 MHz.

Voltage Controlled Wideband Amplifier

The Burr-Brown VCA-610 (Figure 12-5) is an interesting device. It operates up to 30 MHz. It has a control voltage terminal (pin no. 3) that accepts a potential

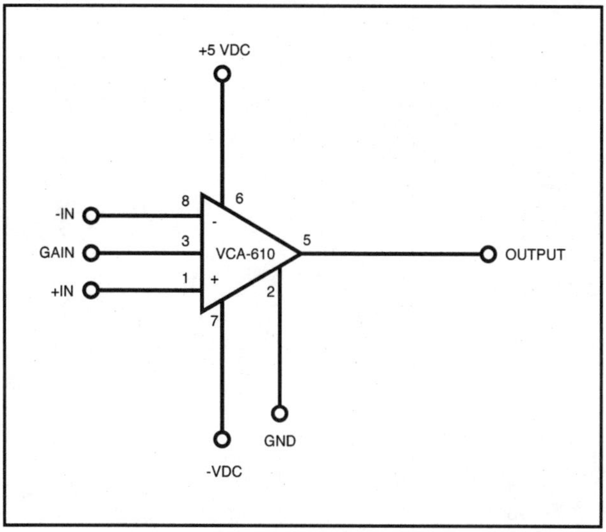

Figure 12-5. Burr-Brown VCA-610 wideband amplifier with voltage controlled gain.

of 0 to -2 volts, and in response provides gains from a -40 dB attenuation to +40 dB amplification (for an overall 80 dB range). Like other wideband products, the design offers a minimum of external components, although bipolar ±5 VDC power supplies are needed.

Figure 12-6 shows a typical circuit for the VCA-610. The +IN input is grounded, and signal is applied to the -IN input. The ±5 VDC power supply lines are bypassed to ground with 0.1 μF capacitors. These capacitors should be selected for high frequency applications, and have no series resonance below 30 MHz. Further, the capacitors must be mounted so that they are as close as possible to the body of the VCA-610.

The control voltage in the circuit of Figure 12-6 is provided through a simple DC voltage divider from the -5 VDC power supply. With the value shown, when the potentiometer wiper is at the top of its resistance element, the voltage at the wiper will be -2 volts.

Because of the voltage controlled feature, the VCA-610 can be used as an automatic gain controlled (AGC) amplifier, in RF and IF circuits, and other applications. One interesting application is in sonar or medical ultrasound. Those applications both fire a pulse of either sound or ultrasound (20 kHz to 10 MHz, but acoustical waves not radio waves). Both sonar and medical ultra-

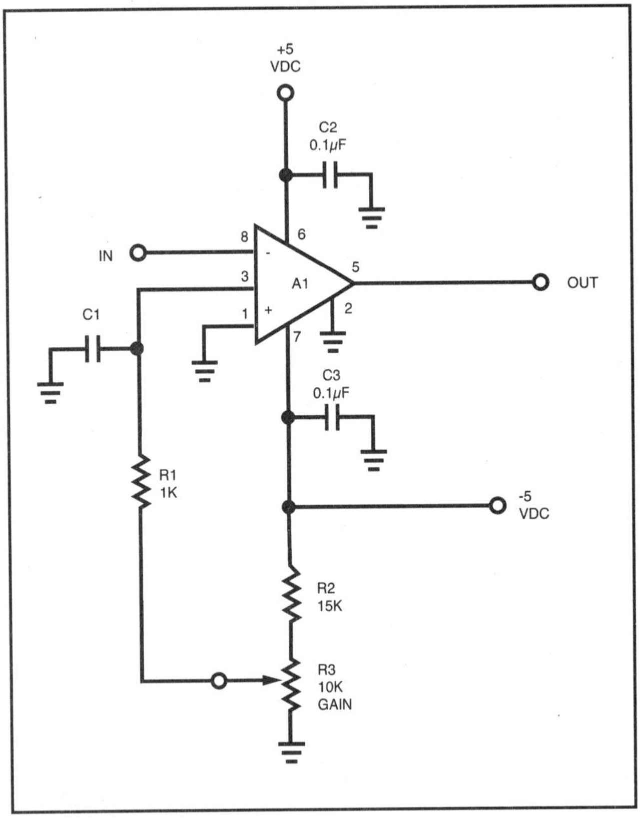

Figure 12-6. Circuit for VCA-610 device provides -40 to +40 dB gain.

sound fire a pulse of energy from a transducer, and then wait for a reflected reply. They need to scale the gain of an amplifier proportional to the time after the pulse is fired. Because of attenuation, the gain of the receiver amplifier must be higher the longer the pulse travels. Otherwise, the gain would either be overwhelmed, or insufficient, depending on how it was set. This feature is sometimes called "time-AGC" or some other name that indicates the gain is a function of time after the pulse is fired.

Wideband Amplifier Construction Hints

Wideband amplifiers give hobbyists and newcomers to the profession of electronics a lot of grief. The reasons for these problems are rooted in the fact that normal "stray" or "distributed" values of capacitance and inductance, found in all circuits (even a straight wire) become significant reactances as frequency increases. When the phase shift caused by such components adds up to 180 degrees in an inverting amplifier, or 360 degrees in a noninverting amplifier, at some frequency where the gain of the amplifier is one or more, then oscillation will result.

Another problem seen in wideband and RF amplifiers is that layout becomes critical. In any amplifier it is good practice to separate the input and output circuitry. The reason is that we need to prevent coupling of output energy back to the input circuit. If this happens, then either degradation or oscillation can occur (depending on phase and gain relationships).

Fortunately, there are some rules (called "heuristics" when we want to be fancy) that permit success most of the time:

1. Keep input and output circuitry separated. If coils or transformers are used, then shielding may be needed to isolate the input and output components.

2. Keep all leads and printed circuit traces as short as possible. When using circuits over about 50 MHz, then use "strip-line" techniques on printed circuit boards.

3. Ensure that input and load impedances are matched to the wideband amplifier (typically 50 ohms, except in TV circuits where 75 ohms is used).

4. Decouple all power supply lines with good, high frequency, capacitors mounted as close as possible to the chip body.

5. Use broad ground tracks on PCBs, and where possible use a ground plane layer on the component side of the board.

6. Keep analog, RF/video (wideband) and digital grounds separated on a printed circuit board except at one common point.

7. Pray.

Chapter 13

Active Analog Filters

A frequency selective filter is a circuit that discriminates against certain frequencies, or bands of frequencies, while passing others. In other words, a filter will pass certain desired frequencies and reject undesired frequencies. Passive filters are made of combinations of passive electronic components such as resistors (R), capacitors (C) and inductors (L). An active filter, on the other hand, is based on an active electronic device such as a transistor or integrated circuit, along with passive components to set the gain, frequency response and other characteristics.

The topic of active filters is a bit too complex for comprehensive coverage in a single chapter, so we will just make a few observations that will allow you a practical basis for building filters. If you want a more comprehensive treatment, then see the references at the end of the chapter.

Filter Frequency Response Characteristics

Filters of all types can be divided into categories based on the passband frequency response. Figure 13-1 shows several popular passband characteristics. In Figure 13-1A we see the low-pass filter (LPF) response. This type of filter

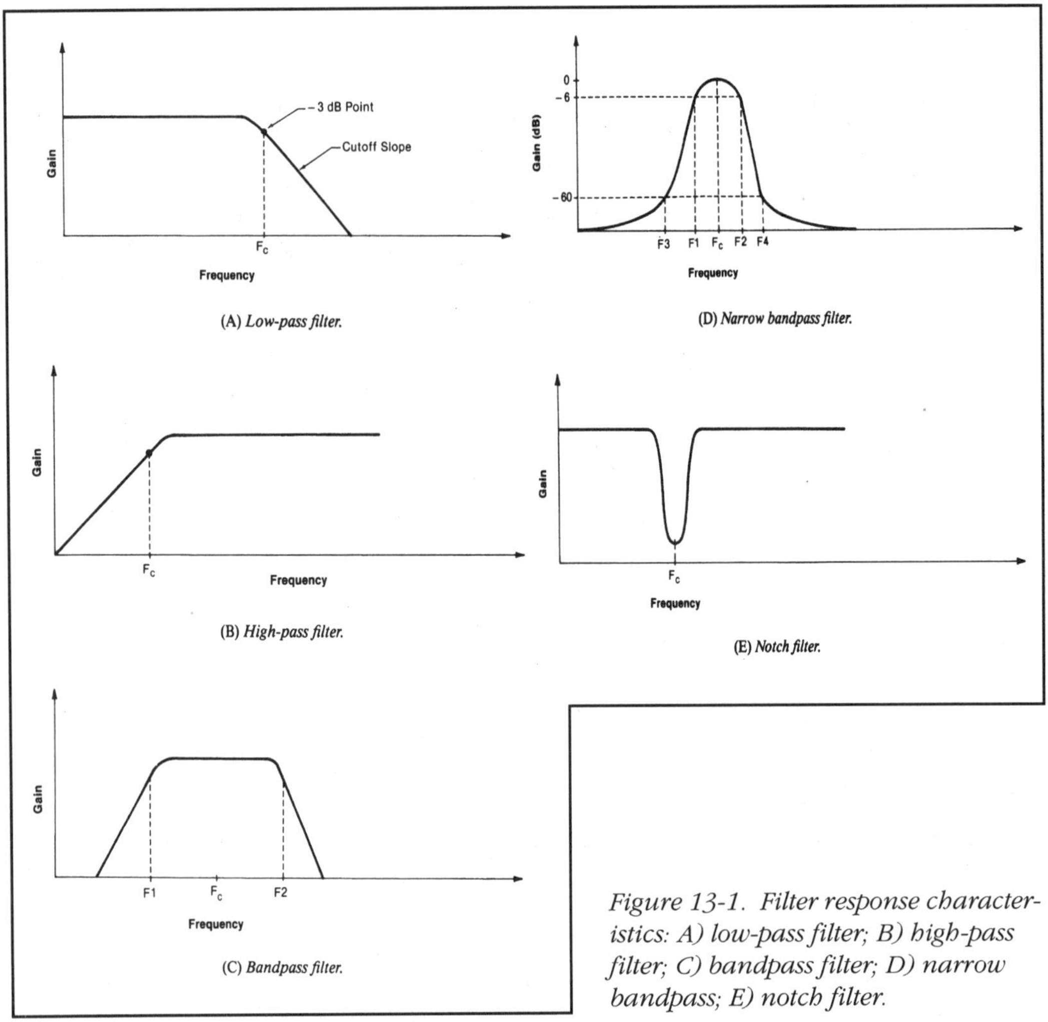

(A) *Low-pass filter.*

(B) *High-pass filter.*

(C) *Bandpass filter.*

(D) *Narrow bandpass filter.*

(E) *Notch filter.*

Figure 13-1. Filter response characteristics: A) low-pass filter; B) high-pass filter; C) bandpass filter; D) narrow bandpass; E) notch filter.

passes all frequencies below a certain critical frequency (F_C). The breakpoint between the passband and the stopband is the point at which the gain of the circuit drops off -3 dB (power) or -6 dB (voltage) from its lower frequency value. Above the critical frequency, the gain falls off at a certain slope. The steepness of the slope is usually given in terms of decibels of gain per octave of frequency (an octave is a 2:1 change in frequency). Alternatively, decibels-per-decade is used (a decade is a 10:1 change in frequency).

The high-pass filter (HPF) response is shown in Figure 13-1B. In this case, the filter circuit will pass all frequencies above the critical frequency, and reject frequencies below that point. The high-pass filter is the inverse of the low-pass filter.

A bandpass filter (BPF) response is one that combines both the LPF and HPF characteristics (Figure 13-1C). In this case, the filter rejects all frequencies below F1 and above F2. These frequencies are the upper and lower -3 dB points in the response; gain is compared with the midband frequency, F_c. The bandwidth of the bandpass filter is the difference between critical frequencies (F2-F1).

The bandpass filter is sometimes characterized by a factor called the figure of merit, or "Q." The higher the Q, the narrower the frequency response. So as Q goes up, F2-F1 gets smaller. The Q of any bandpass circuit is defined as:

- $$Q = \frac{F_c}{F2 - F1} \qquad \text{eq. (13-1)}$$

Equation (13-1) is sometimes written in a form that recognizes that F2-F1 is the bandwidth (BW) of the circuit:

- $$Q = \frac{F_c}{BW} \qquad \text{eq. (13-2)}$$

The response of Figure 13-1C is of the low-Q type because it has a wide bandwidth. A higher-Q circuit has a narrower bandpass, as in Figure 13-1D. It admits a smaller range of acceptable frequencies.

A notch filter, also called a stopband filter, will pass all frequencies except for those that are clustered around a critical frequency (Figure 13-1E). The notch filter is used to take out a single unwanted frequency. A common use for notch filters is to remove 60 Hz interference signals from electronic circuits. This job is done a lot in scientific and medical instruments. Amateur radio operators sometimes use a notch filter circuit when receiving Morse code continuous wave (CW) signals in order to notch out nearby interfering signals. The active analog notch filter is often a cheap substitute for a high-priced narrow band mechanical or quartz bandpass IF amplifier filter. In that application, the operator will tune the receiver to make the desired signal have a tone of, say, 400 Hz. Interfering signals further away, but still within the receiver's

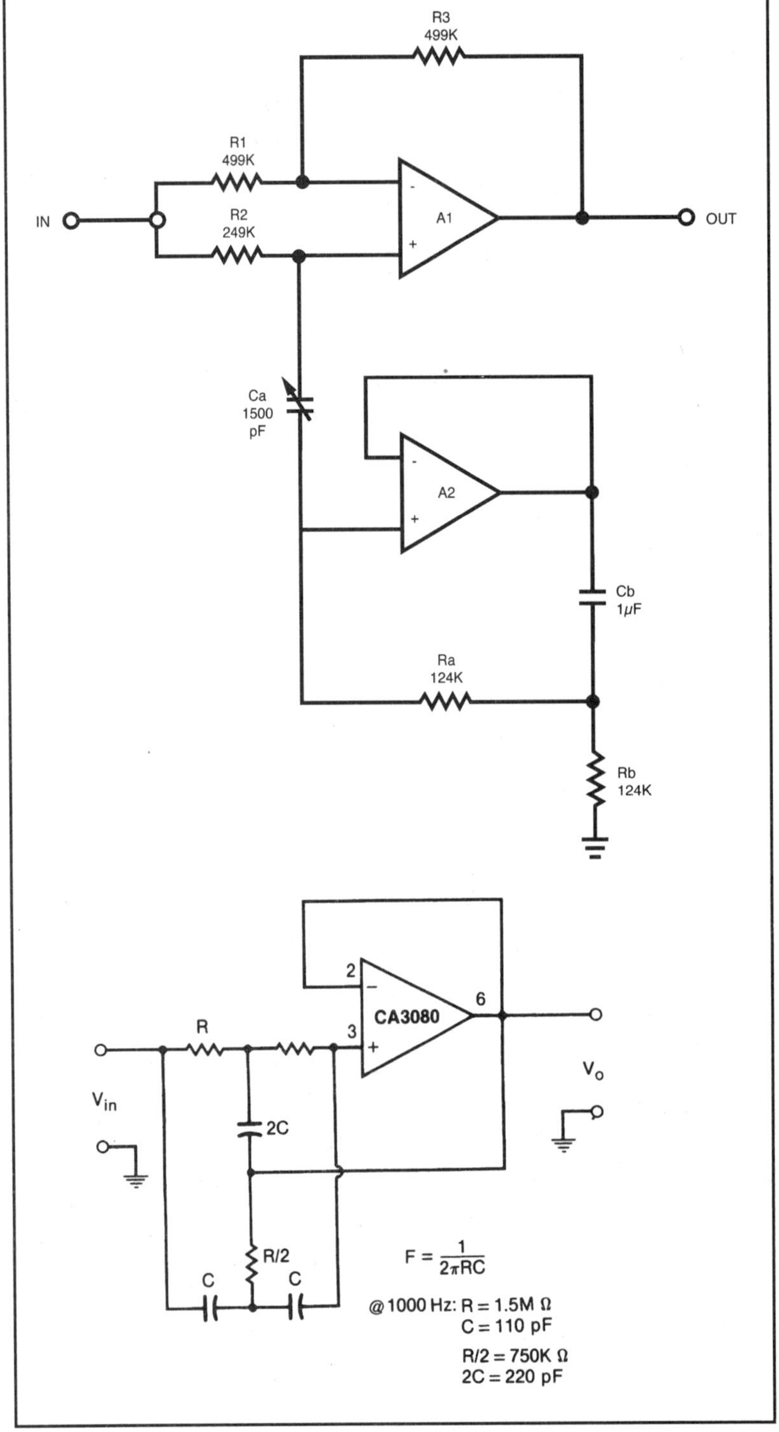

Figure 13-2. A) standard notch filter circuit; B) variable "gyrator" notch filter circuit.

passband, may have a frequency of, say, 970 Hz. These signals can be eliminated by placing a 970 Hz notch filter in the audio amplifier chain. An example of a notch filter centered near 1,000 Hz is shown in Figure 13-2A. If you desire a 60 Hz notch, then replace the resistors and capacitors with values determined from:

- $$F_c = \frac{1{,}000{,}000}{2\pi RC}$$ *eq. (13-3)*

— where:

F is the critical frequency in hertz (Hz).
R is the resistance in ohms (Ω).
C is the capacitance in microfarads (μF).

For the 60 Hz model, select a capacitor value in the 0.005 μF to 0.05 μF range (0.01 μF is a good initial choice).

The notch filter of Figure 13-2A is difficult to make variable. Either the two resistors marked "R" must be replaced with a dual potentiometer (and the two pots must track each other as the shaft is rotated), or a dual-section or split-stator variable capacitor must be provided for the capacitors marked "C." Both have been done, but the tracking problem and the mechanical problems sometimes make one wonder if it's worth it. A better solution for many applications is the variable "gyrator" circuit of Figure 13-2B. In this circuit, a pair of operational amplifiers are used. The notch frequency of this circuit is normalized for 60 Hz, and is found from:

- $$F_c = \frac{1}{2\pi\sqrt{R_a R_b C_a C_b}}$$ *eq. (13-4)*

— or, if Ra = Rb, and Ca = Cb:

- $$F_c = \frac{1}{2\pi R_a C_a}$$ *eq. (13-5)*

In actual practice, one would probably want to keep the resistors R_a and R_b, and capacitor C_b, at some value at, or close to, those shown for 60 Hz, and make the capacitor C_a an air variable type to "tune" the actual frequency.

High-Pass and Low-Pass Filter Designs

Very simple active analog filters are shown in Figure 13-3. A low-pass filter is shown in Figure 13-3A and a high-pass filter is shown in Figure 13-3B. These designs are really little more than buffered versions of simple RC passive filters. The values shown in Figure 13-3 are normalized for 1 kHz. We can find the actual values (R' and C') required for any other frequencies by dividing the normalized 1 kHz values by the desired cut-off frequency expressed in kilohertz.

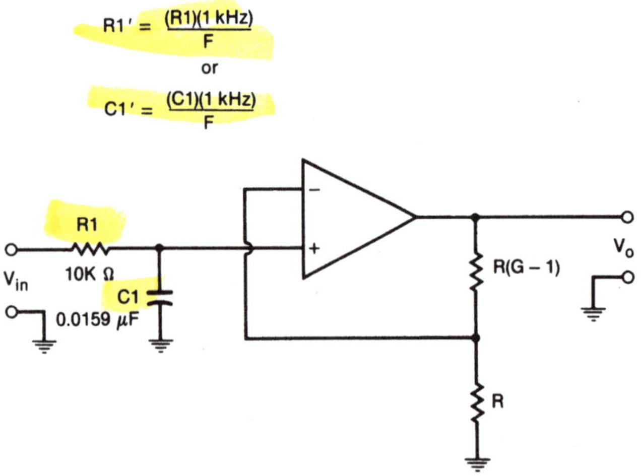

(A) ***A 1-kHz normalized low-pass filter.***

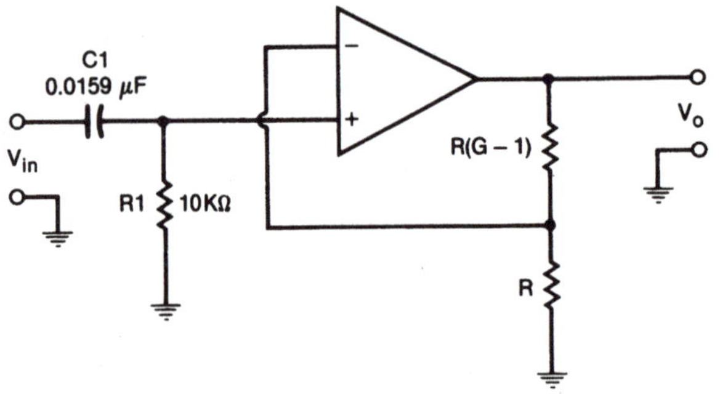

Figure 13-3. A) Simple R-C low-pass filter normalized for 1 kHz; B) simple R-C high-pass filter normalized for 1 KHz.

Example

Select values for either LPF or HPF at a frequency of 60 Hz (note: 60 Hz = 0.06 kHz).

Solution

- $C1' = \frac{R1 \times 1KHz}{F_{desired}}$ *eq. (13-6)*
- $C1' = \frac{R1 \times 1KHz}{0.06KHz} = 0.265\mu F$ *eq. (13-7)*

We must leave one of the values alone, and then calculate the other. In the example above, 0.265 µF is not a standard value, but 0.27 µF is standard. There are three courses of action open to use here. First, just stick in a 0.27 µF unit and live with the results; two, hand-select a 0.265 µF unit from a collection of 0.27 µF units (tolerances are such that there is sure to be one); and three parallel connect several lesser value capacitors until the actual total sum value is 0.265 µF. Custom selection of capacitors was, at one time, not too easy for hobbyists, but today digital capacitance meters, or digital multimeters with a capacitance function, are widely available and easily obtained at low cost.

A model for the simplistic design of high-pass and low-pass filters is shown in Figure 13-4. In this generalized case, we have a series of four reactances, labelled Z1 through Z4. These components are replaced with either resistances or capacitances. Examples are shown in Figure 13-5 (low-pass filter case) and Figure 13-6 (high-pass filter case). Again, the values are normalized for 1 kHz, and the actual values are selected using the equations shown below:

Low-Pass Case

- $R1' = R2' = \frac{(10K)(1KHz)}{F_{KHz}}$ *eq. (13-8)*
- $C1' = \frac{(0.0122\mu F)(1KHz)}{F_{KHz}}$ *eq. (13-9)*
- $C2' = 2C1'$ *eq. (13-10)*

High-Pass Case

- $R1' = \frac{(7.07K)(1KHz)}{F_{KHz}}$ *eq. (13-11)*
- $R2' = 2R1'$ *eq. (13-12)*
- $C1' = C2' = \frac{(0.0159\mu F)(1KHz)}{F_{KHz}}$ *eq. (13-13)*

A different form of active filter circuit, called the multiple feedback path circuit, is shown in Figure 13-7. The low-pass version is shown in Figure 13-7A, while the high-pass version is shown in Figure 13-7B. The values are normalized for 1 kHz, and we find the actual required values in the manner described above. Change either the capacitor or the resistor value, but not both.

A practical circuit for a multiple function filter is shown in Figure 13-8. Again, the values are normalized for 1 kHz. This filter is interesting because it provides both low-pass and bandpass functions. Output V_{O1} is the bandpass output, while output V_{O2} is the low-pass output.

The filter in Figure 13-8 is close to what is called a state variable filter (SVF). The values shown are normalized for 1 kHz. As in the previous cases, for an SVF, divide the R and C values by the desired frequency in kilohertz in order find the values for any particular project.

One decision that you need to make before calculating the resistance and capacitance value is the Q of the filter (see definition above). You must deter-

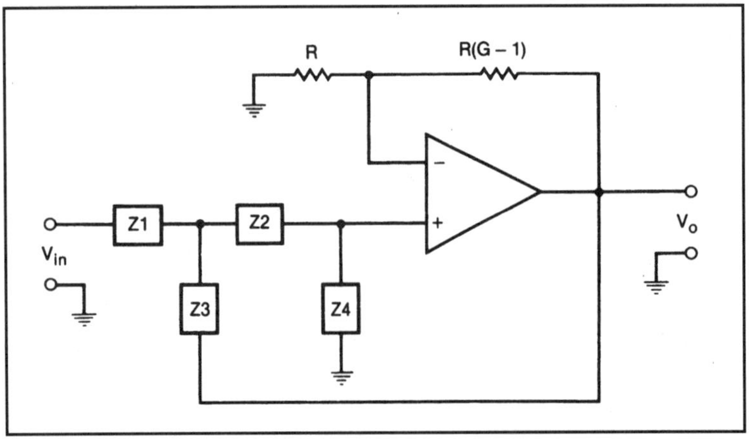

Figure 13-4. Active analog filter model circuit.

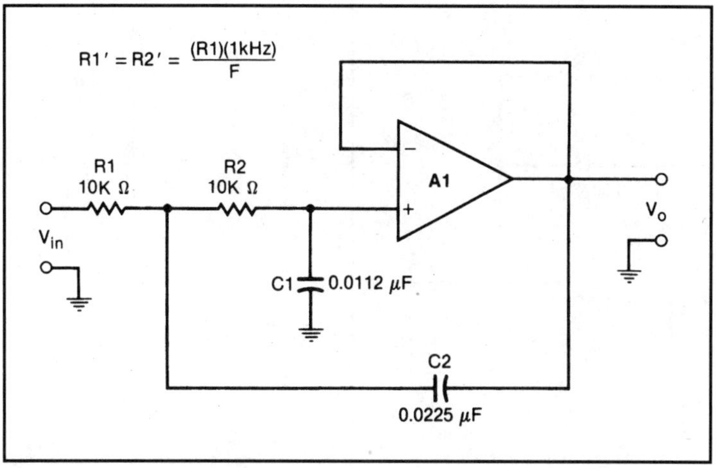

Figure 13-5. Low-pass filter normalized for 1 KHz.

mine from the application what center frequency and bandwidth is needed on the bandpass filter function, and from that information calculate the Q.

Example

What Q is required of a filter to pass a 2,000 Hz bandwidth center around 10 kHz? (Note: 2,000 Hz = 2 kHz)

- $$Q = \frac{F_c}{BW} = \frac{10KHz}{2KHz} = 5 \qquad eq.\ (13\text{-}14)$$

Once the Q is known, we can calculate the filter damping factor α, which is the reciprocal of Q:

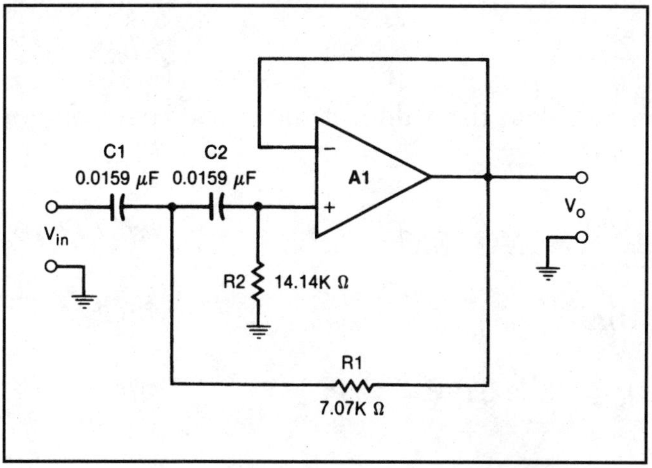

Figure 13-6. High-pass filter normalized for 1 KHz.

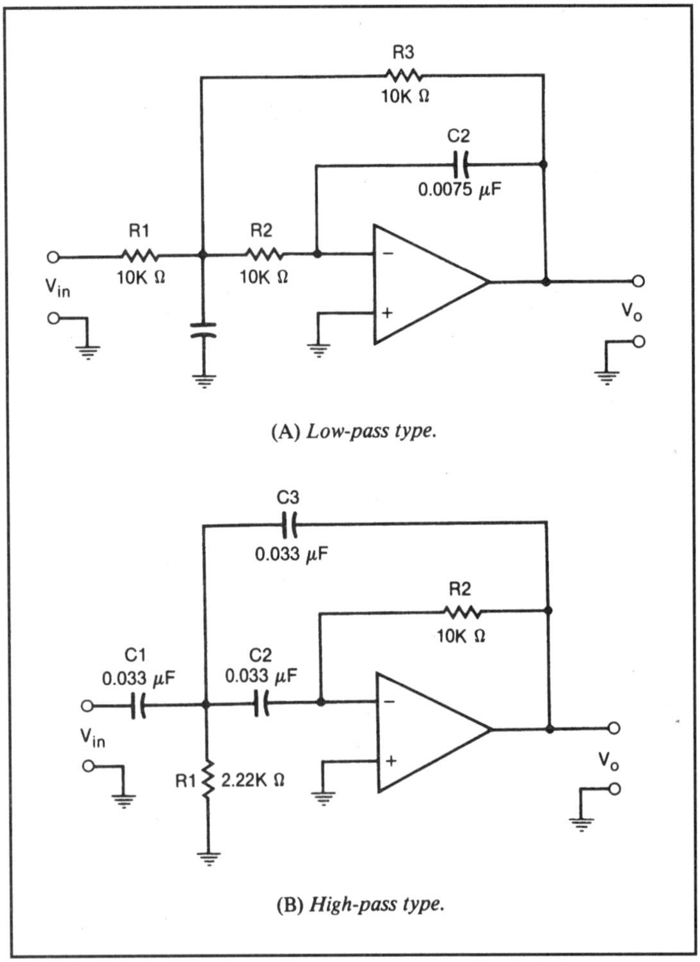

Figure 13-7. Multiple feedback path active analog filter.

- $\alpha = \frac{1}{Q} = 0.2$ *eq. (13-15)*

Now that α is known, we can calculate the value of the rest of the components. Again, for the 10 kHz filter:

- $R' = \frac{(10K)(1KHz)}{10KHz} = 1K@10KHz$ *eq. (13-16)*
- $R2 = R = 10K$ *eq. (13-17)*
- $R1 = \frac{3R}{\alpha} - R = \frac{(3)(1K)}{0.2} - 1K = 14K$ *eq. (13-18)*

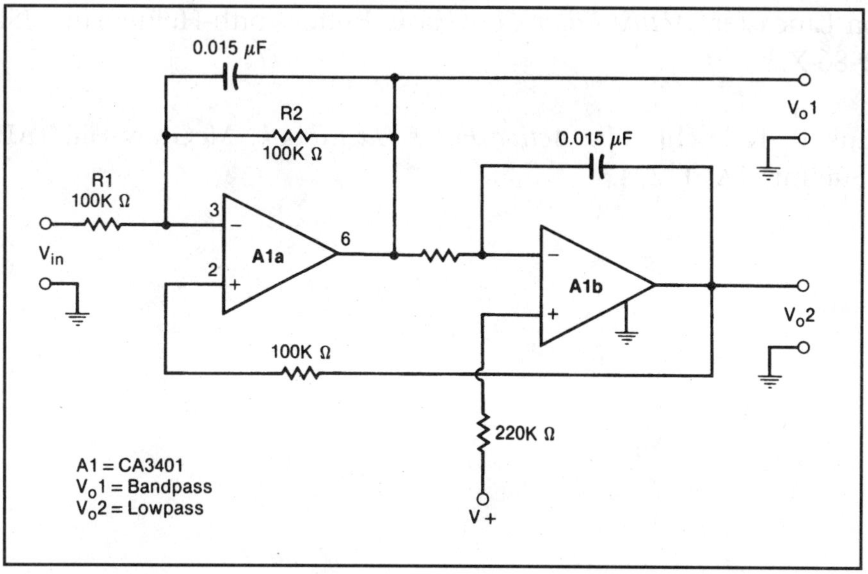

Figure 13-8. Bandpass filter/low-pass filter normalized for 1 KHz.

- $$C' = \frac{(0.0159\mu F)(1KHz)}{10KHz} = 0.00159\mu F \qquad eq.\ (13\text{-}19)$$

In Chapter 15 you will find a type of circuit called a Miller integrator. Because of certain problems inherent in op amp active integrators, you must use good quality capacitors for C, and the op amps used should be types with a low input bias current, and low output offset voltage. Otherwise, performance will suffer when the bias current or other cause of output offset voltages causes the capacitors to charge. Typical low-cost example include the CA-3140 device, or the "two-fer" version of the same thing, the CA-3240. Many of the special function IC filters sold on the market today are actually state variable filters, but with the op amps and resistors built into the DIP package. Some of them have small variable capacitors built in as well.

For Further Reading

1. Howard M. Berlin, *Design of Active Filters, with Experiments.* Howard W. Sams. Out of print, but still seen in libraries.

2. Don Lancaster, *Active Filter Cookbook*. Butterworth-Heineman. ISBN #0-7506-2986-X.

3. Frank P. Tedeschi, *The Active Filter Handbook*. McGraw-Hill/TAB, Blue Ridge Summit, PA, 17294.

Chapter 14

Waveform Generators and Timer Circuits

Almost everyone who dabbles in electronics eventually needs to build an oscillator, multivibrator or digital clock circuit. These circuits are used to produce a variety of electronic waveforms that are needed in many different circuits and applications. A simple oscillator, for example, might be used in a radio transmitter or receiver. Such circuits are generally considered sine wave generators, even though the word "oscillator" is correctly applied to circuits that produce other waveforms. The multivibrator could produce square waves, triangles or other waveforms. Similarly, a digital clock is a special case of the multivibrator and/or oscillator that is used in digital logic and computer circuits.

In general, let's keep our jargon straight by adopting the term "oscillator" to cover all three cases, including astable multivibrators and digital clocks. The term *oscillator* can be defined as a circuit that will produce a periodic waveform (i.e., one that repeats itself). The output waveform can be a sine wave, square wave, triangle, sawtooth, pulse or other wave shape. The important part is that it is periodic.

There are two forms of oscillator circuit: relaxation oscillators and feedback oscillators. Some relaxation oscillators use one of several available "negative resistance" devices (e.g., tunnel diode). Such devices operate according to Ohm's law at certain potentials, and opposite Ohm's law at other potentials. Other relaxation oscillators use devices that pass little or no current at voltages below some threshold, and pass a large current at voltages above the threshold. Examples of these devices are neon glow-lamps and unijunction transistors (UJT).

Feedback oscillators use an active device as an amplifier, and then provide feedback in a manner that produces regeneration instead of degeneration. These circuits account for a large number of oscillators used in practical electronic circuits. In this series of articles, we will discuss both forms of oscillator.

Relaxation Oscillators

There are two main devices used in practical relaxation oscillators: neon glow-lamps and UJTs. There are also some circuits that are based on SCR-like devices, but these are not common. We also have a class of relaxation oscillators that are based on IC devices such as voltage comparators, integrators and so forth.

Figure 14-1A shows a simple relaxation oscillator based on the neon glow-lamp. The reason for including it in a book on ICs is to demonstrate the principle of the relaxation oscillator. The neon glow-lamp has a low-pressure inert gas inside of a glass envelope that also contains a pair of electrodes. When the potential across the electrodes exceeds the ionization potential (V_t in Figure 14-1B) of the inert gas, the gas gives off light. Popular types of neon glow lamp include the NE-2 and the NE-51. For these lamps, the threshold potential is somewhere between 40 and 80 volts, although the lamp will maintain its ionized state (and continue to glow) at a somewhat lower "holding potential" (V_H).

The circuit for the relaxation oscillator is shown in Figure 14-1A, while its operating waveforms are shown in Figure 14-1B. The frequency of oscillation is set by the ionization and holding potentials, acting in concert with the RC

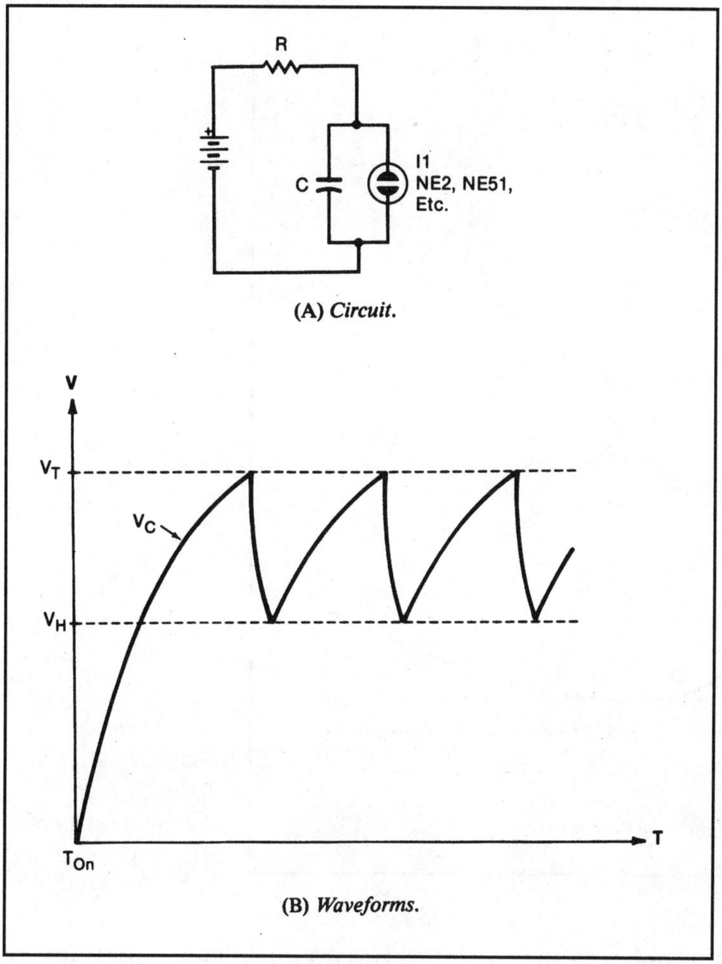

Figure 14-1. Simple relaxation oscillator: A) circuit; B) waveform.

time-constant of the resistor and capacitor. The resistor is connected in series with the lamp, while the capacitor is connected in parallel with the lamp.

We must keep in mind the two states of the neon lamp. When the lamp is not ionized, it conducts no current. This is the situation at all potentials below V_t. At voltages above the threshold, the lamp resistance suddenly drops to nearly nothing (in fact, the lamp can blow if the resistor is too low value). When the circuit is turned on, the capacitor will begin to charge and the capacitor voltage (V_c) will begin to build up. When the voltage across the capacitor reaches the threshold potential, the lamp ionizes and takes on a very low resistance. Since this low resistance is across the capacitor, it will discharge the capacitor.

(A) *Symbols.*

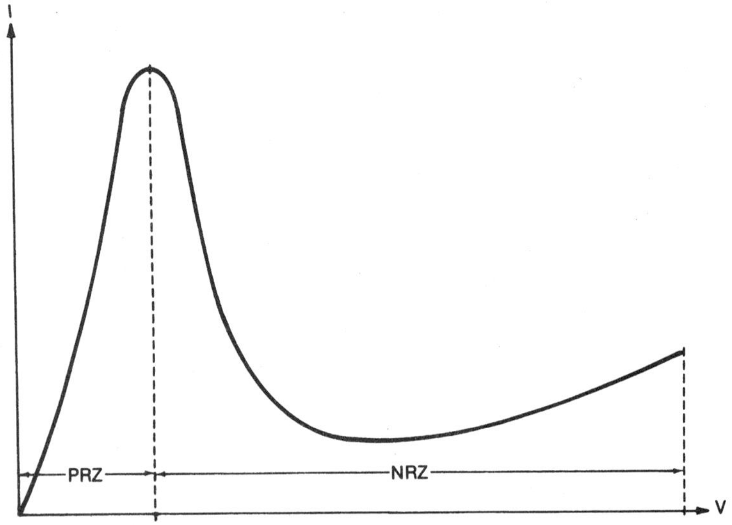

(B) *Current versus voltage curve.*

Figure 14-2. Tunnel diode device: A) symbols; B) I-vs-V curve.

But the discharge only continues until V_c reaches the threshold potential. When this potential is reached, the lamp goes out and reverts to its high resistance state. The cycle continues, with the capacitor voltage varying between threshold and holding potentials. The frequency of oscillation is determined by the difference between these potentials and the values of R and C.

Negative Resistance Devices

There are several semiconductor devices on the market that have a so-called negative resistance property. What does this mean? Regular, or "positive resistance," operates according to Ohm's law. When the voltage across the circuit is increased, the current in the device increases proportionally; when the volt-

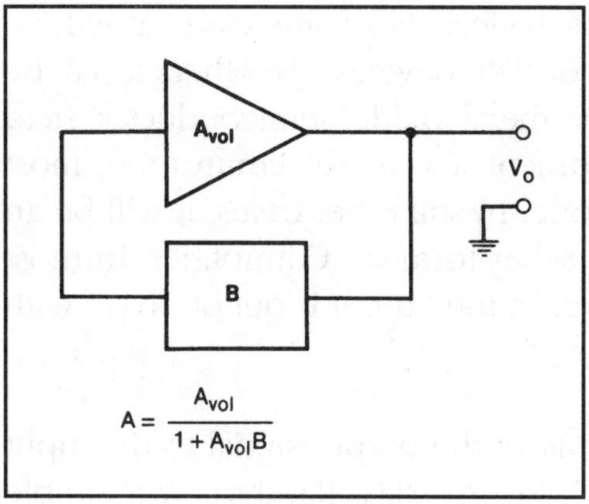

Figure 14-3. Equivalent circuit for a feedback oscillator.

age is decreased, the current decreases also. In a negative resistance device, however, exactly the opposite is found. An increase in voltage will produce a decrease in current, and vice versa.

An example of a negative resistance device is the tunnel diode (also called an Esaki diode after the inventor), the symbols for which are shown in Figure 14-2A. The I-vs-V curve for the tunnel diode is shown in Figure 14-2B. In the positive resistance zone (PRZ), the tunnel diode operates much like other "Ohm's law" devices. However, in the negative resistance zone (NRZ), it behaves just the opposite from Ohm's law!

Relaxation oscillators are simple, and are good circuits for students and beginning hobbyists to build (if only to learn how they work). Most other applications, however, will use one or another feedback oscillator circuits.

Feedback Oscillators

A feedback oscillator consists of an amplifier and a feedback network (see Figure 14-3). It is called a "feedback oscillator" because the output signal of the amplifier is fed back to the amplifier's own input by way of the feedback network. We call the gain of the amplifier A_{vol}, and the transfer function of the oscillator B. The overall gain of this circuit is given by the relationship:

$$A_v = \frac{A_{vol}}{1 + A_{vol}\beta} \qquad \text{eq. (14-1)}$$

The amplifier can be any of many different devices. For some cases, it will be a common-emitter bipolar transistor (NPN or PNP devices). In others it will be a junction field effect transistor (JFET) or metal oxide semiconductor field effect transistor (MOSFET). In older equipment it was a vacuum tube, most often operated in the common-cathode mode. In still other cases, it will be an integrated circuit operational amplifier, or other form of IC amplifier. In most cases, the amplifier will be an inverting type, so the output is out of phase with the input by 180 degrees.

The feedback network will provide a sample of the output signal to the input of the amplifier. This network may be LC, RC, or RLC. Rarely is it a simple

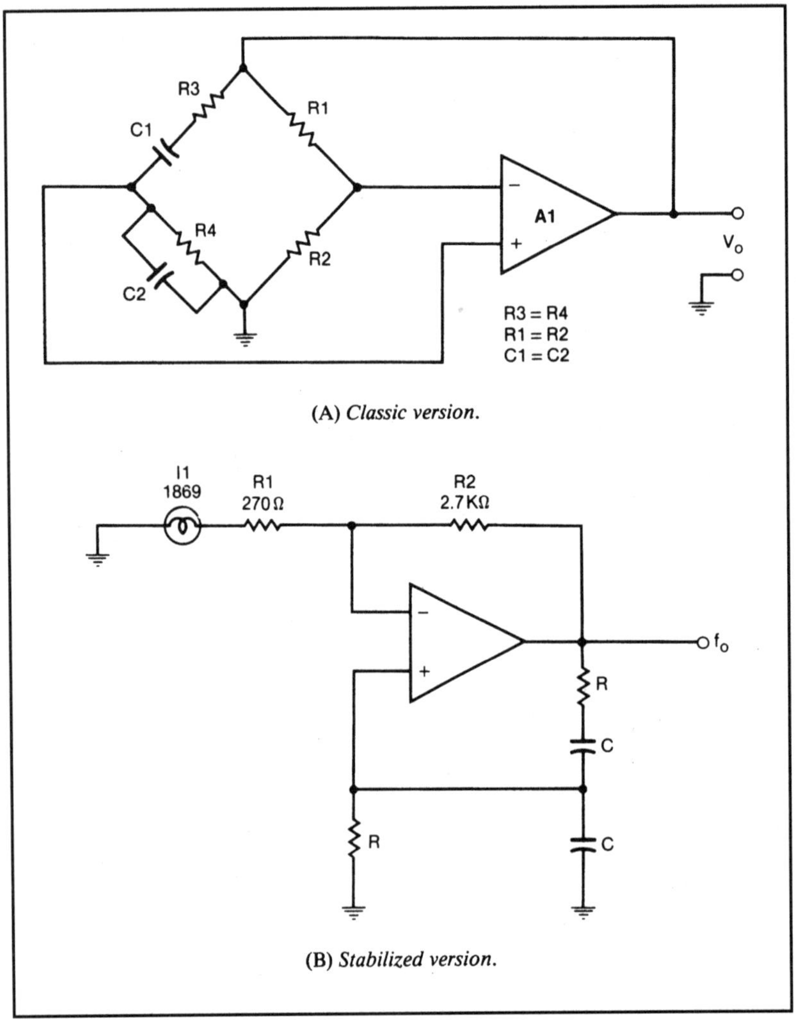

Figure 14-4. Wein bridge oscillator: A) classic circuit; B) amplitude stabilized version.

resistance network, because such networks are not frequency selective. They will produce oscillation, however, but at some frequency that is not controlled. The problem is that there are distributed ("stray") circuit capacitances in the circuit, and these act with the resistance to produce an RC time-constant at some unspecified frequency — rarely a frequency that you want!

This circuit will oscillate if two conditions are met: (1) the loop gain (as above) is greater than unity, and (2) the feedback signal is in-phase with the input signal. Since the typical amplifier inverts the signal (180 degrees), and since we need a total phase shift of 360 degrees (in order to have an in-phase feedback signal), the feedback network must provide an additional phase shift of 180 degrees. If the network is designed to produce this required phase shift at only one frequency (as in an LC or RC network), then the circuit will oscillate at that frequency.

In the rest of this chapter, we will discuss several forms of oscillator circuit in which the active element is either an analog or digital integrated circuit.

Sine Wave Oscillators

Sine wave generators produce an output signal that is sinusoidal in nature. Such a signal is very pure, and if a perfect waveform, will contain only a fundamental frequency with no harmonics. The harmonics of non-sinusoidal waveforms are what gives it a characteristic shape. The active element in all of these circuits is the ordinary operational amplifier.

Wein Bridge Oscillators

Figures 14-4A and 14-4B are examples of Wein bridge oscillators. The elementary form is shown in Figure 14-4A. The operating frequency in the general case is:

- $$F_{Hz} = \frac{1{,}000{,}000}{2\pi\sqrt{R3R1C1C2}} \qquad \text{eq. (14-2)}$$

— where:

F_{Hz} is frequency in hertz (Hz).

All resistors in ohms (Ω).
All capacitors in microfarads (μF).

— or, in the special case where R1 = R2, R3 = R4, and C1 = C2:

- $$F_{Hz.} = \frac{1{,}000{,}000}{2\pi R3C1}$$ *eq. (14-3)*

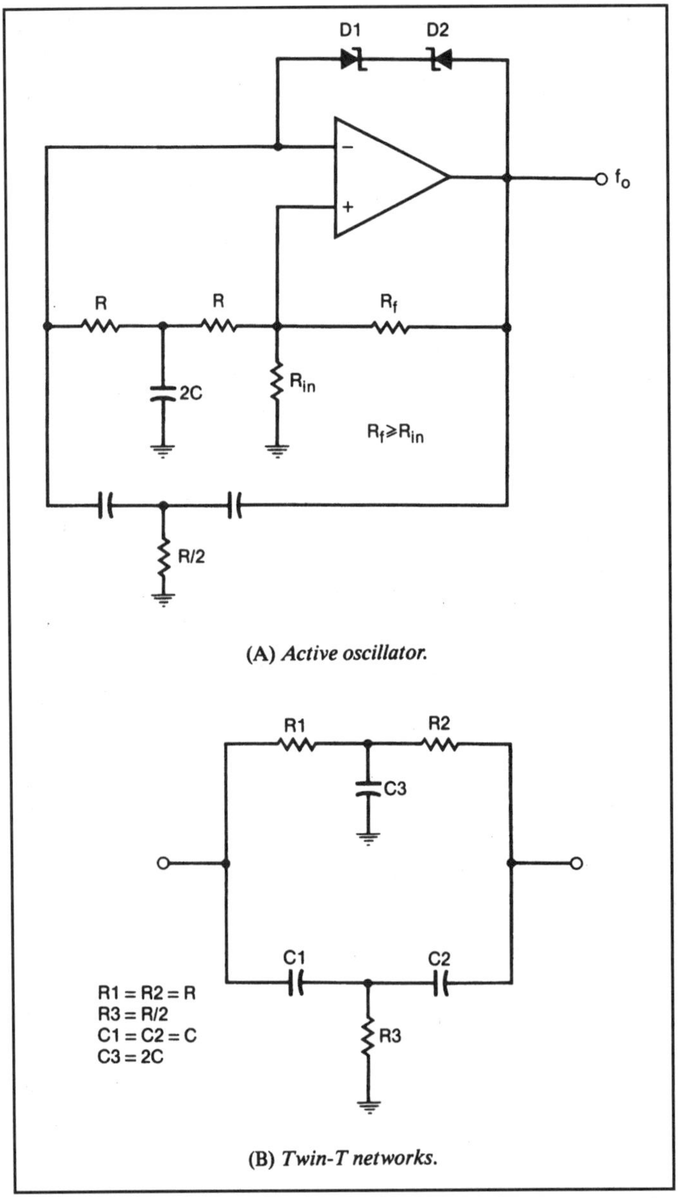

Figure 14-5. Twin-T oscillator A) circuit; B) Twin-tee network.

A problem with this oscillator is the fact that the output amplitude varies, especially as the frequency is changed. We can overcome that problem in part with the rearranged circuit in Figure 14-4B. In this circuit we place an incandescent lamp in one of the bridge arms to overcome the variable amplitude problem. As the amplitude of the output voltage increases, so the does the current in lamp I1 (a #1869 lamp is used for I1). This reduces its resistance, thereby lowering the gain of the amplifier. This relationship is easier to see in the redrawn circuit. The operating frequency is found from:

- $$F_{Hz} = \frac{1{,}000{,}000}{2\pi 11 RC}$$ *eq. (14-4)*

— where:

FHz is in hertz (Hz).
R is in ohms (Ω).
C is in microfarads (uF).

Twin-Tee Oscillators

A twin-tee network is a two-branch RC circuit as shown in Figure 14-5. There are two tee-networks in this circuit: R1/R2/C3 and R3/C1/C2. The required 180 degrees phase shift is created by this circuit at the frequency given by:

- $$F_{Hz} = \frac{1{,}000{,}000}{2\pi RC}$$ *eq. (14-5)*

— where all terms are as previously defined.

Diodes D1 and D2 are used to clip the output if the gain is too high, but that event should be limited by adjusting the ratio R_f/R_{in}.

Bridged-Tee Oscillators

A variant of the tee oscillator is the bridged-tee of Figure 14-6. In this circuit, one of the tee circuits is replaced with a single resistor bridging the remaining tee network. A variant on the circuit reverses the roles of all the elements. In that circuit, R2 and R3 are capacitors, while C1 and C2 are resistors — the principles are the same, however.

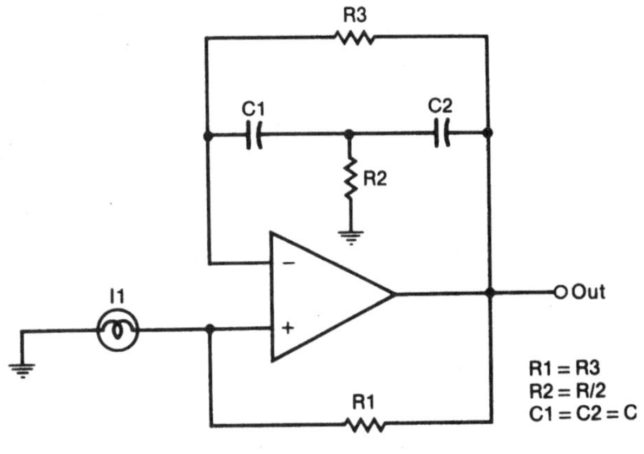

Fig. 18-6. Bridged-T oscillator.

Figure 14-6. Bridged Tee oscillator circuit.

Phase Shift Oscillators

Figure 14-7A shows a phase shift network consisting of three resistors and three capacitors. Each resistor-capacitor combination (R1/C1, R2/C2 and R3/C3) produces a 60-degree phase shift at the "resonant" frequency. When combined, the overall phase shift is 180 degrees — which is exactly as required for oscillation.

Figure 14-7B shows an oscillator circuit based on the phase shift network. The active element is an operational amplifier, while R1-R3/C1-C3 form the feedback network. The operating frequency is found from:

$$F_{Hz} = \frac{1{,}000{,}000}{2\pi\sqrt{6}\,RC} \qquad \text{eq. (14-6)}$$

— or, when all of the constants are collected into one:

$$F_{Hz} = \frac{64{,}975}{RC} \qquad \text{eq. (14-7)}$$

— where all terms are as previously defined (Provided that R1 = R2 = R3 = R, and C1 = C2 = C3 = C).

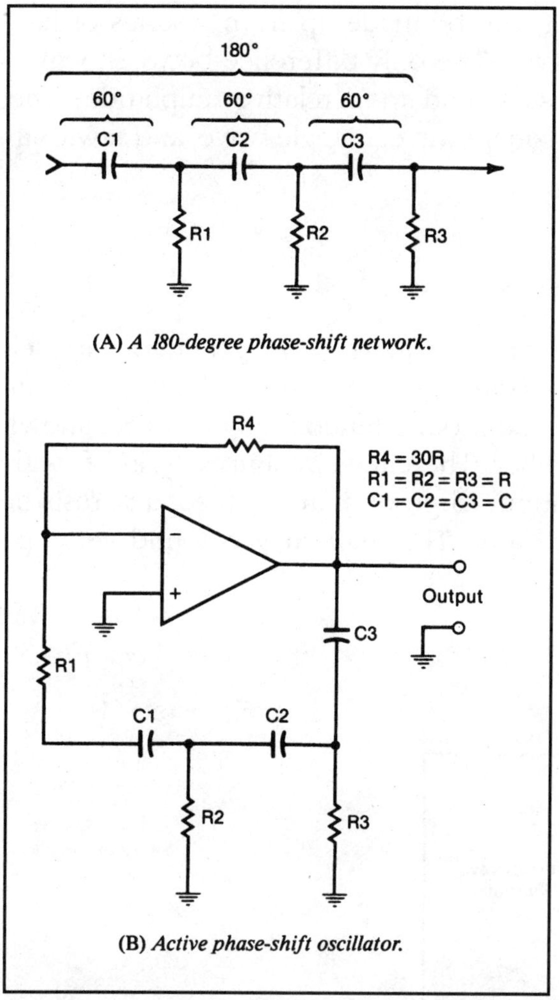

Figure 14-7. R-C phase shift oscillator: A) A 180-degree R-C phase shift network; B) Active phase-shift oscillator based on an op amp.

The overall gain of the circuit of Figure 14-7B is the sum of the forward gain of the amplifier (-R4/R1) and the "gain" of the feedback network. The feedback network gain is a loss of about 1/29 at the resonant frequency, so the ratio R4/R1 must be at least 30R in order to make up for the loss and permit oscillation (recall the criteria for feedback oscillators above).

Non-Sinusoidal Waveform Generators

All non-sinusoidal waveforms contain a fundamental frequency plus a collection of sine and cosine harmonics of the fundamental. In fact, all mathematical functions, whether waveforms or not, can be made up from a series of harmonically-related sine and cosine waves. The only difference between waveforms is the specific harmonics present, and their relative amplitudes and phases. In this section we will discuss square wave, triangle wave, and sawtooth wave generators.

Square Wave Generators

Figure 14-8 shows two examples of simple square wave generator circuits. Although these circuits produce waveforms with poor rise and fall time characteristics, they are simple enough to merit our attention. Figure 14-8A shows a circuit based on the CA-3048 amplifier. The circuit is simplicity itself, with only a capacitor from the inverting input to ground, and a feedback resistor from the output to the noninverting input. The frequency is found very approximately from:

$$F_{Hz} = \frac{1,000,000}{2\pi RC} \qquad \text{eq. (14-8)}$$

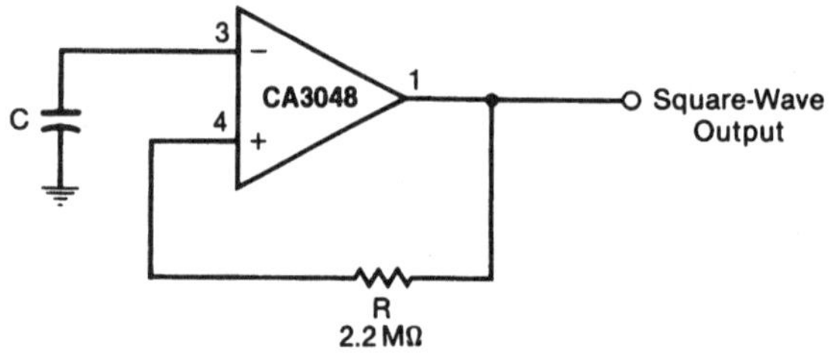

(A) *CA3048.*

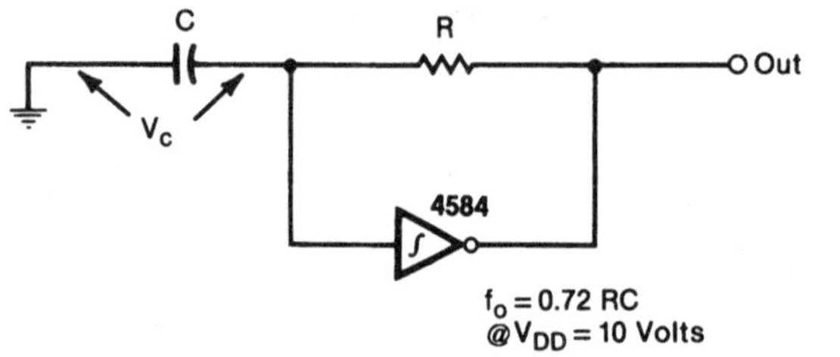

(B) *4584 CMOS.*

Figure 14-8. Square wave oscillator.

The second circuit, Figure 14-8B, is based on the CMOS Schmitt trigger device. A Schmitt trigger output snaps HIGH when the input voltage crosses a specified threshold potential in a positive-going direction, and drops LOW again only after the input voltage crosses a second threshold potential in a negative-going direction. The inverting Schmitt trigger offers a HIGH output until the first threshold is crossed, and a LOW after the threshold potential is reached.

At turn-on, capacitor C is discharged, so the voltage at the input of the 4584 is zero. This potential is below the triggering threshold, so the output remains HIGH. The HIGH output charges capacitor C through resistor R. When the capacitor voltage (V_c) reaches the Schmitt trigger point, the output of the 4584 drops LOW. Now that the output is LOW, the capacitor begins to discharge through R, and when it reaches the second trip point the output snaps HIGH again. The HIGH/LOW seesaw continues as long as power is applied.

An operational amplifier square wave generator circuit is shown in Figure 14-9. The description of the operation of this circuit is similar to that for Figure 14-8B, except that an operational amplifier is used. The "trip points" are set by the feedback resistors, R1 and R2. The general case gives us a 50-percent duty cycle square wave with a period of:

$$T_{sec} = 2R3C1LOG\left(\frac{1+\beta}{1-\beta}\right) \qquad eq.\ (14\text{-}9)$$

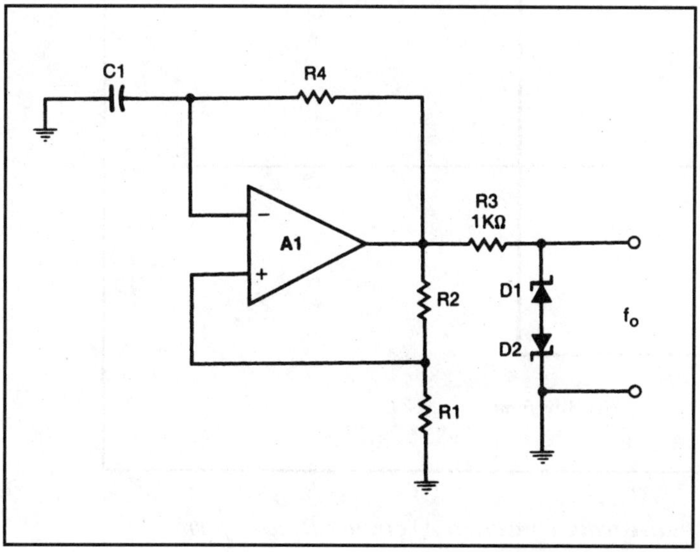

Figure 14-9. Op-amp square wave oscillator.

— where:

T_{SEC} period in seconds (s).
R3 is in ohms (Ω).
C1 is in farads (F)
$\beta = (R1/[R1 + R2])$

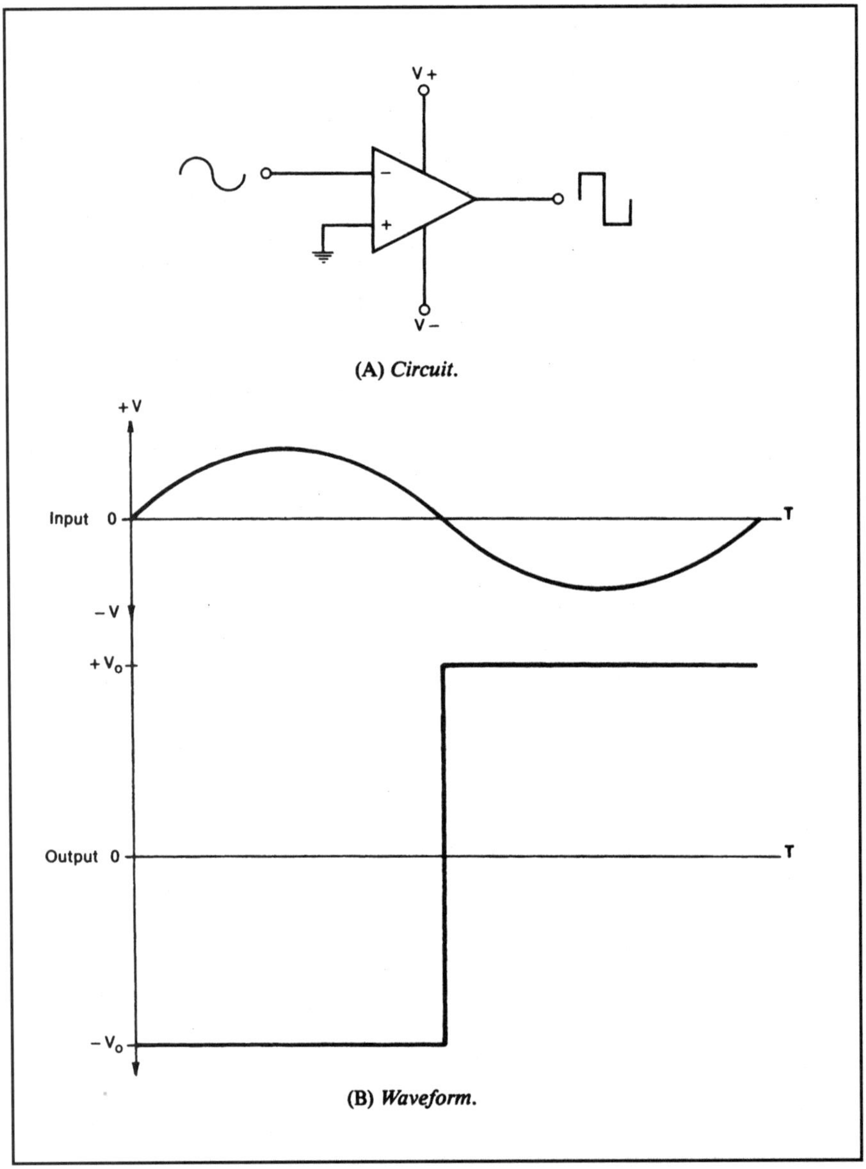

Figure 14-10. Sine-to-square wave converter: A) circuit; B) waveform.

A special case for the above equation if B = 0.46 (which we can force by making R2 equal to 1.17R1 (example: R1 = 15k, and R2 = 18k comes close to this ideal using standard value resistors). If this criteria is met, then:

- $$T_{sec} = 2R3C1$$ *eq. (14-10)*

In either event, the operating frequency is the reciprocal of the period:

- $$F = \frac{1}{T}$$ *eq. (14-11)*

The diodes (D1 and D2) in Figure 14-9 are used to limit the output voltage to a value less than V- and V+. Operational amplifiers tend to saturate when the output hits the supply rail, at least for a short period. As a result, the rise and fall times are adversely affected.

Square Waves from Sine Waves

Figure 14-10 shows the method for converting sine waves to square waves. The circuit is shown in Figure 14-10A, while the waveforms are shown in Figure 14-10B. The circuit (Figure 14-10A) is an operational amplifier connected as a comparator. Because the op amp has no negative feedback path, the gain is very high; with typical op amps gains of 20,000 to 2,000,000 are found. Thus, a voltage difference across the input terminals of only a few millivolts will saturate the outputs. From this behavior, we can understand the operation of the circuit, and the waveform in Figure 14-10B.

The input waveform is a sine wave. Because the noninverting input is grounded (Figure 14-10A), the output of the op amp is zero only when the input signal voltage is also zero. When the sine wave is positive, the output signal will be at $-V_o$; when the sine wave is negative, the output signal will be at $+V_o$. The output signal will be a square wave at the sine wave frequency, with an amplitude of $[(+V_o)-(-V_o)]$. An output level control, variable gain post-amplifier stage or an attenuator network can be used to reduce the signal to a usable level.

Earlier we discussed textbook integrator circuits. We learned that these circuits can be used as (1) low-pass filters, (2) to generate quadrature sine wave sig-

266 / Chapter 18

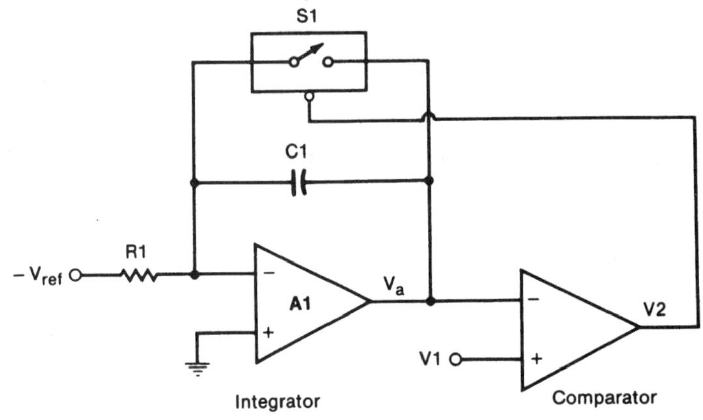

Figure 14-11. Trial sawtooth oscillator.

nals, and (3) to generate triangle waves out of square waves. We will look at some practical circuits that use the integrator.

Sawtooth Generator Circuit

A sawtooth waveform is formed when a linear voltage ramp is abruptly terminated, and the voltage drops back to zero. There are several methods for generating a sawtooth. Figure 14-11 shows one attempt at designing a sawtooth circuit. It consists of an integrator (A1) followed by a voltage comparator (A2); the output of the comparator drives the control input on an electronic switch (which is active-LOW). At turn-on, the charge in capacitor C1 is zero; so voltage V_a is zero. Because voltage V1 is positive, the output of the comparator (V2) is positive. Under this circumstance the control line of S1 is inactive, so S1 is open.

At turn-on, the stable reference voltage -Vref causes the output of the integrator to ramp upwards. At the point where V_a = V1, the output of A2 drops LOW, forcing S1 to close. This discharges C1, forcing the output voltage (V_a) close to zero.

Figure 14-12 is the classic op amp sawtooth generator circuit. The ramp generator circuit is a Miller integrator formed with A1, C1 and R3 (being driven by

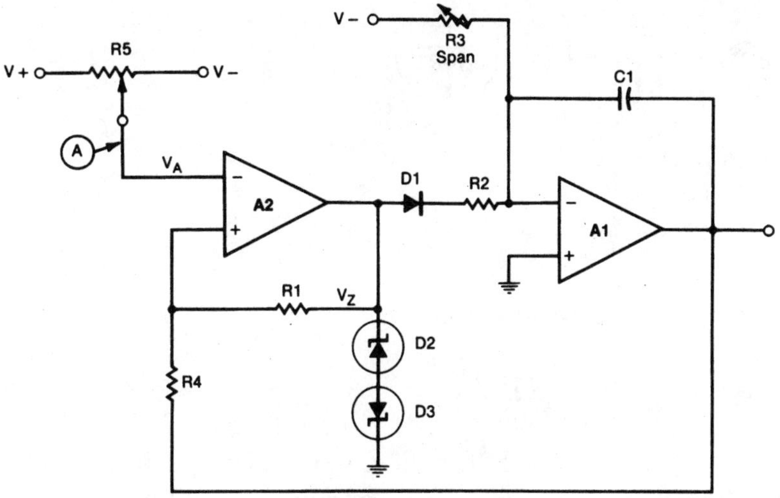

R3>>R2
R1 = R4

$$T = \frac{2V_Z R3C1}{V-}$$

Figure 14-12. Classic op-amp sawtooth oscillator.

V-). The output voltage ramps up until it reaches the threshold of comparator A2. That comparator uses positive feedback and a reference voltage V_{ref} provided by a potentiometer. The trip threshold is V_{ref} + V1 (which is set to 0.7 volts greater than the zener voltage of D1/D2, assuming that these diodes are identical). When the output voltage hits the threshold voltage, the comparator output snaps positive and forward biases diode D3. If resistor R2 is very much less than R3, then C1 will discharge very rapidly, resulting in the classical sawtooth wave shape. The reset time will be very much shorter than the period if R2 <<< R3.

Another means of generating a sawtooth waveform is to use a special function generator chip, such as the Exar XR-2206 (which is available from Jameco and others). Although in another circuit the XR-2206 will generate sine/square wave/triangle waveforms, in the circuit of Figure 14-13A the chip generates a sawtooth and a short duty factor (DF) pulse.

The frequency of this sawtooth generator circuit is set by resistors R1 and R2, plus capacitor C1:

- $$F_{Hz} = \left(\frac{2}{C1}\right)\left(\frac{1}{R1+R2}\right)$$ *eq. (14-12)*

— where:

F_{Hz} is in hertz.
C1 is in farads.
R1 and R2 are in ohms.

The duty factor is found from:

- $$D.F. = \frac{R1}{R1+R2}$$ *eq. (14-13)*

Jameco makes a circuit board "function generator" kit that creates the sine wave, square wave and triangle wave signals. It can be easily modified for sawtooth applications. The sawtooth and pulse output waveforms of Figure 14-13A are shown in Figure 14-13B.

The same Exar XR-2206 device can also be used for generating sine wave and square wave signals. Figure 14-14 is the circuit for this application. Switch S1 determines whether the signal at pin no. 2 is a sine wave or triangle waveform. This circuit is similar, but not identical, to the Jameco function generator circuit. The operating frequency is:

- $$F_{Hz} = \frac{1}{RC1}$$ *eq. (14-14)*

The amplitude and DC offset of the output waveform is adjusted using trimmer potentiometers R6 and R7.

Digital Waveform Generators

Modern digital electronics technology can be used to make purely analog waveforms. This type of circuit is the basis for many electronic music projects today. Figure 14-15 shows generally how this job is accomplished. The heart of the circuit is a digital-to-analog converter (DAC). The output current of the DAC0806 is converted to a voltage by amplifier A1, and then filtered to re-

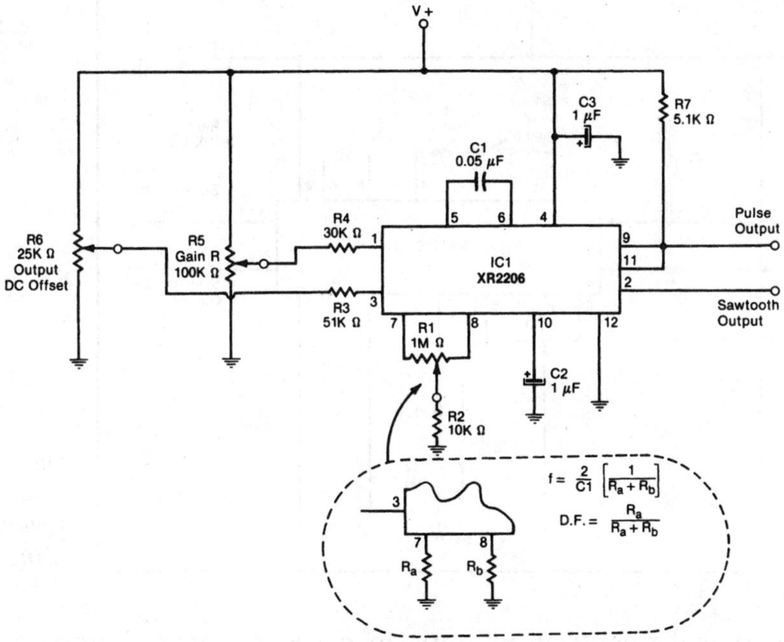

(A) *Circuit.*

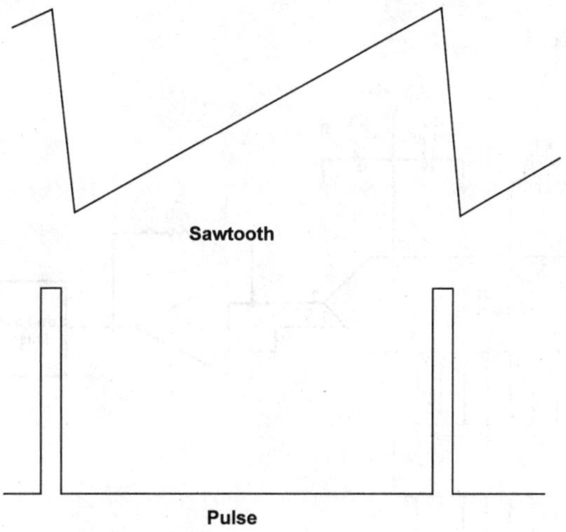

Figure 14-13. XR-2206 waveform generator IC: A) circuit; B) waveforms.

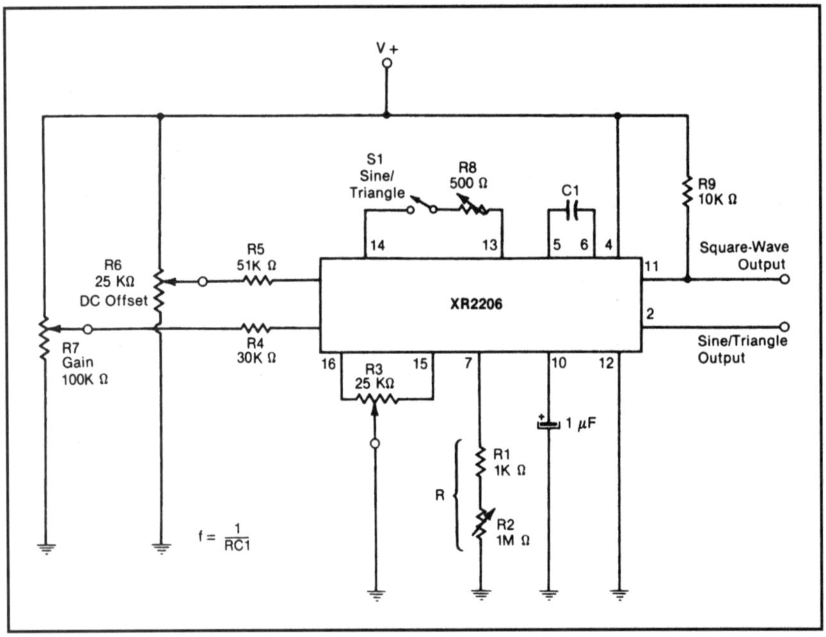

Figure 14-14. Sine/triangle/ square generator using XR-2206.

moved the digital quantization steps in a second-order low-pass filter. We select the waveform by changing the binary input to the DAC.

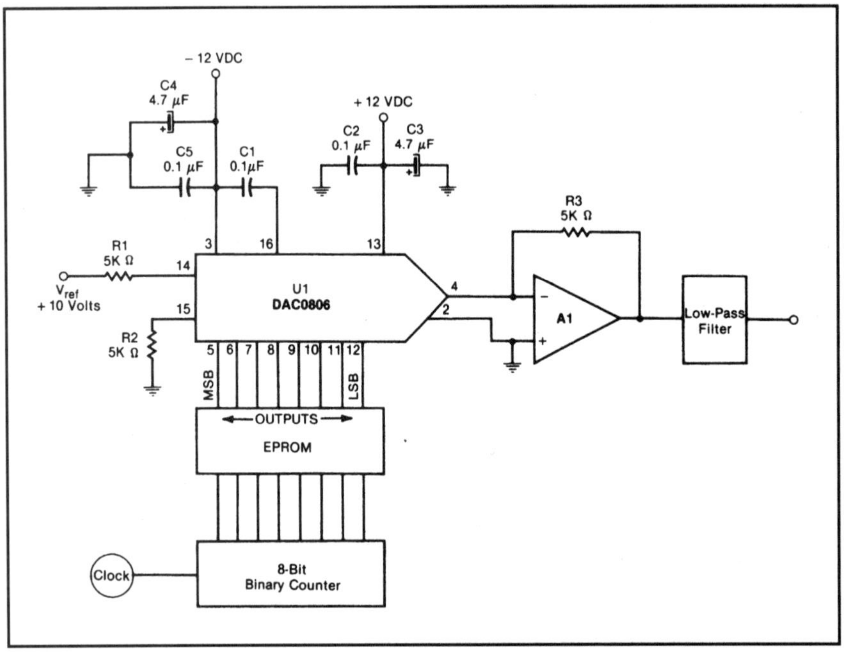

Figure 14-15. Digital sawtooth oscillator.

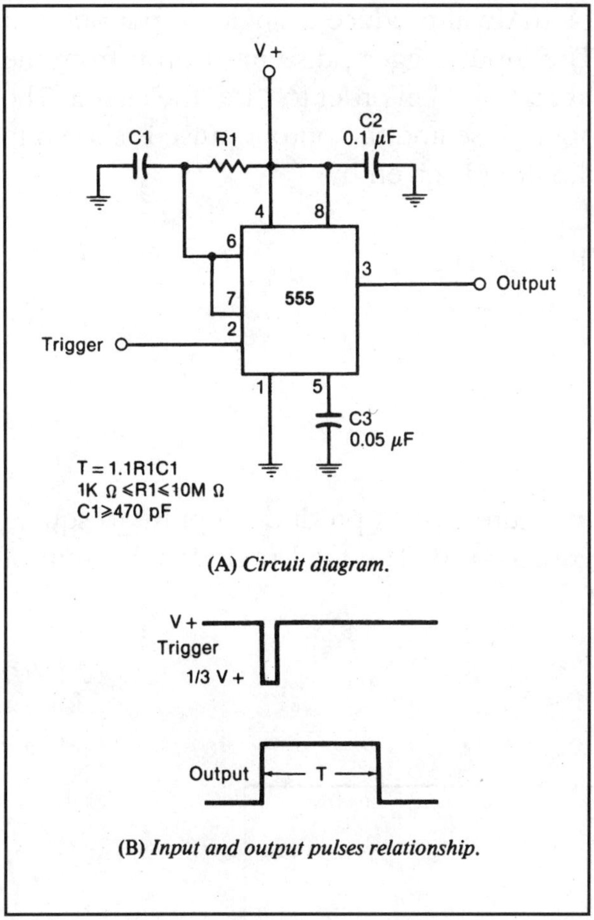

(A) *Circuit diagram.*

(B) *Input and output pulses relationship.*

Figure 14-16. LM-555 monostable timer circuit.

If we need a sawtooth, then it is merely necessary to connect the eight-bit DAC inputs to the outputs of an eight-bit binary counter. As long as the clock runs, the output will oscillate back and forth as a sawtooth with a frequency equal to the clock frequency divided by 256. The binary patterns for other waveforms can be stored in an erasable programmable read-only memory (EPROM).

Timer Circuits

A timer chip is a device that can be used as a monostable multivibrator (one-shot) or astable multivibrator, and is timed with a resistor-capacitor network, as shown in Figures 14-16 and 14-17 (monostable and astable, respectively).

The monostable circuit of Figure 14-16A will produce a single output pulse of duration T for each input pulse. The input trigger pulse must drop from the HIGH (close to V+) to a value less than (V+)/3 in order to "fire" the circuit. The relationship between the trigger input pulse and the output pulse is shown in Figure 14-16B. The output pulse duration is given by:

- $$T_{\text{sec}} = 1.1R1C1 \qquad \textit{eq. (14-15)}$$

— where:

T_{SEC} is in seconds (s).
R1 is in ohms (Ω).
C1 is in farads (F).

The astable configuration shown in Figure 14-17A produces a pulse to square wave-like waveform shown in Figure 14-17B. The total period is the sum of HIGH time (T1) and LOW time (T2):

- $$T = t1 + t2 \qquad \textit{eq. (14-16)}$$

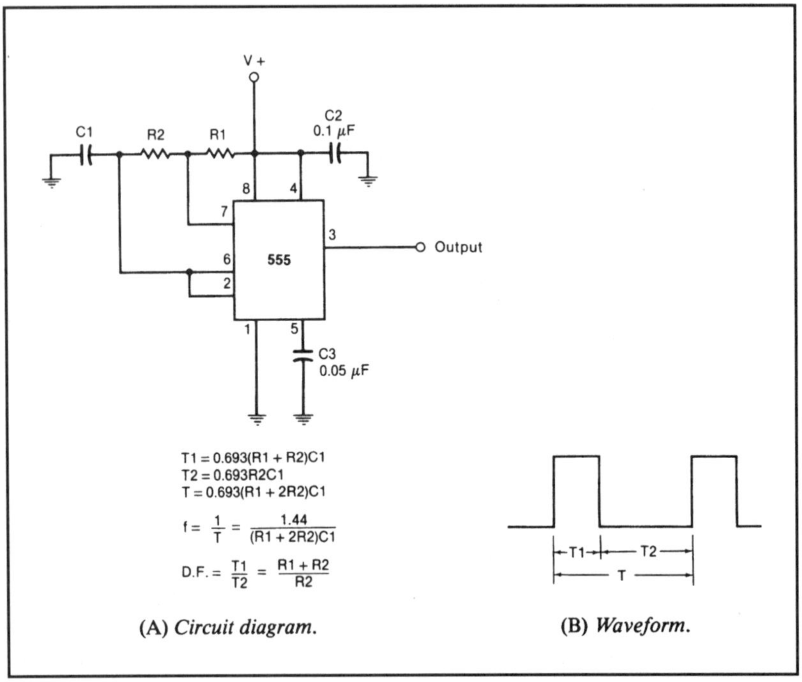

(A) *Circuit diagram.* (B) *Waveform.*

Figure 14-17. LM-555 astable timer/ square wave generator circuit.

— where:

- $t1 = 0.693(R1 + R2)C1$ *eq. (14-17)*

— and:

- $t2 = 0.693R2C1$ *eq. (14-18)*

Therefore:

- $T = 0.693(R1 + 2R2)C1$ *eq. (14-19)*

The frequency is found from:

- $F_{Hz} = \frac{1.44}{(R1+2R2)C}$ *eq. (14-20)*

Conclusion

Waveform generator circuits serve a wide variety of uses, and can be built easily using the guidelines given in this chapter.

Chapter 15

Instrumentation and Other Circuits

"Electronic instrumentation" circuits represent a very large class, depending upon how you define "instrumentation." In this chapter, therefore, we will take a look at several miscellaneous circuits that are generally useful in constructing electronic instruments. We will also take a look at the general class of circuits called proximity detectors. The reason for this is to demonstrate the concept of transduction by using a nonstandard example.

Keeping the Circuits Stable

Although we discussed stability op amp problems in an earlier chapter, I want to reopen the issue briefly in order to drive the point home. In preparing this chapter I had to breadboard several circuits (integrators and differentiators) to check out the principles presented. Ordinarily, common 741 operational amplifiers are used for casual projects because those devices are both low-cost and well-behaved. Unfortunately, the "well-behaved" point is due to their being "unconditionally stable." In other words, the frequency response is rolled off so much that there is little chance of oscillation. That fact makes it easy to become slack in such matters as power supply bypassing. The 741 will frequently operate normally with the power supply leads unbypassed; but op

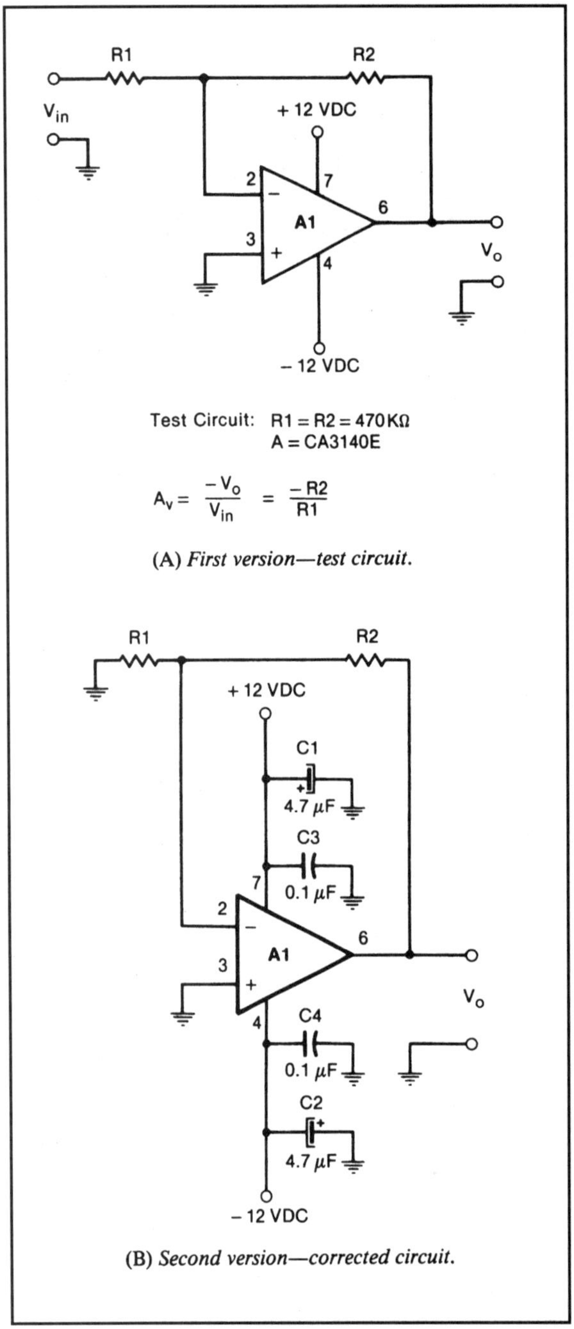

(A) *First version—test circuit.*

(B) *Second version—corrected circuit.*

Figure 15-1. Inverting follower circuits: A) test circuit; B) corrected version.

amps with higher frequency responses (like the CA-3140 BiMOS device) will become unstable and oscillate if the power supply is unbypassed.

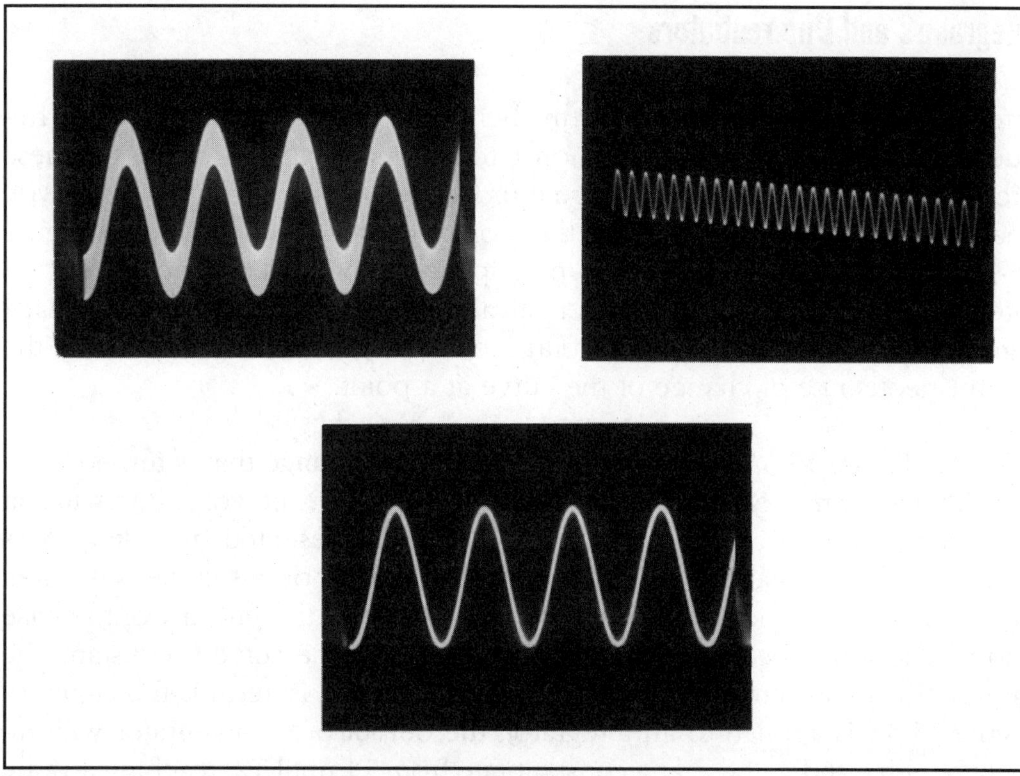

Figure 15-2. Waveforms of circuit of Fig. 15-1: A) 400 Hz sine wave with superimposed RF oscillation; B) oscillation waveform; C) corrected waveform.

Figure 15-1A shows the first circuit that I connected. Because the differentiator and integrator circuits being checked are based on the inverting follower, I first connected an inverting follower with a gain of one (R1 = R2 = 470 K) to test the op amp from my junk box and the connections. The idea was to then replace the appropriate resistors with the correct integrator or differentiator components. When a 400 hertz sine wave signal was applied to the input, the oscilloscope showed the waveform in Figure 15-2A. Note the thickening of the trace: the fuzziness indicates oscillation at high frequency. Expanding the trace (shown in Figure 15-2B) demonstrated that the oscillation was at a frequency somewhat higher than 20 kilohertz. The CA-3140 operational amplifier has a higher gain-bandwidth product than the old-fashioned 741, and required the bypass capacitors shown in Figure 15-1B. When I connected those capacitors into the circuit, the oscilloscope showed the trace in Figure 15-2C — which is the correct trace.

Integrators and Differentiators

Integration and differentiation are mathematical processes that are used frequently in electronic instrumentation circuits. Although the details of these processes are from calculus, and are thus beyond the scope of this book, you need to know a few details about them on at least a descriptive level. Figure 15-3 demonstrates some of these principles. Keep in mind two facts: the integral is the area under a mathematical curve. For a time-varying voltage signal, the integral is the time-average of the signal. The derivative is the instantaneous rate of change of the curve at a point.

Consider Figure 15-3A. In this case, we have a DC voltage that is turned on at time T1 and turned off at time T2. Between T1 and T2 the voltage remains at a constant level, V1. The mathematical "curve" represented by voltage V is rather simple because V = V1 at all times. The integral of this curve is the area under the curve, which for this simple case is merely the product of the base and height of the rectangle: V1 x (T2-T1). Because the curve is constant, the rate of change is zero, so the derivative of the curve is zero. If the signal of Figure 15-3A is applied to an integrator, the output of the integrator will initially be zero, and will rise at a constant rate from T1 until T2, reaching a value dependent upon the values of V1 and (T2-T1).

A second case is shown in Figure 15-3B. In this case, the voltage signal is a constantly rising voltage. Because the rate of change is constant, we can measure the voltage change between any two time points and find the rate of change. For the situation shown in Figure 15-3B:

$$\frac{\Delta V}{\Delta T} = \frac{V2-V1}{T2-T1} \qquad \text{eq. (15-1)}$$

— where:

ΔV represents the time change of voltage, V.

Δ represents the change in time, T.

In many cases, when the interval is small enough, the Greek letter "delta" (Δ) is replaced with a lowercase "d." That convention makes the expression above read:

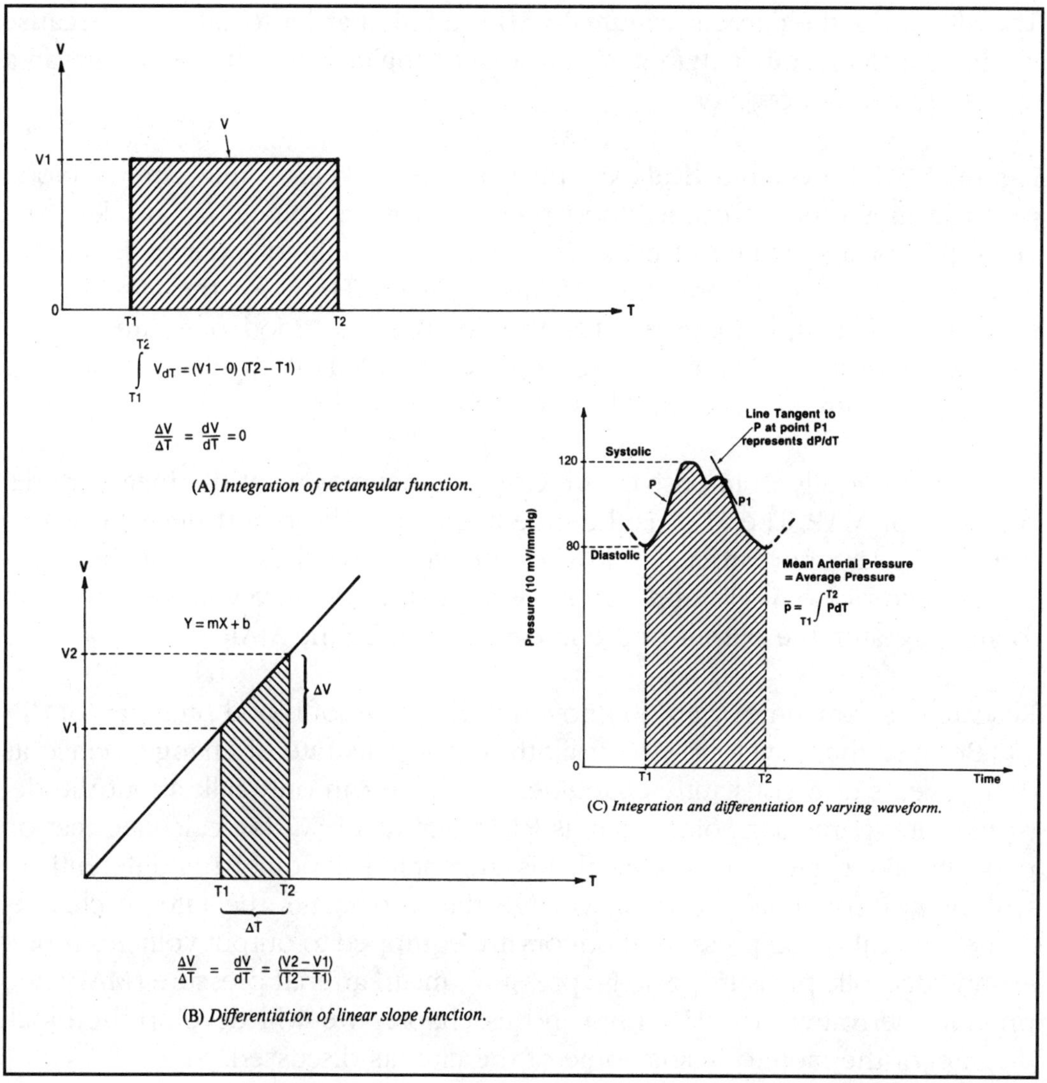

Figure 15-3. Integration and differentiation: A) of rectangular function; B) of linear slope function; C) of varying waveform.

- $$\frac{dV}{dT} = \frac{V2-V1}{T2-T1} \qquad eq.\ (15\text{-}2)$$

Some purists would find the above distinction between "delta" and "d" somewhat lacking, but that fact does not harm our discussion here.

The integral of the curve in Figure 15-3B is a little harder to calculate because we have a figure other than a square or rectangle. It is still merely the area under the curve, however.

Figure 15-3C shows a practical example from medical electronics. Arterial blood pressure monitors represent blood pressure with a voltage. In modern devices, the voltage analog of pressure is usually 10 millivolts per millimeter of mercury (i.e., Torr) of pressure (10 mV/mmHg). Thus, for the case shown (which is where this overweight author wishes his blood pressure was), a systolic pressure of 120 mmHg is represented by 1200 mV, while the 80 mmHg diastolic pressure is represented by 800 mV.

Doctors, especially in intensive care cases, sometimes want the mean arterial pressure, or MAP. The MAP is the time average of the blood pressure waveform (P) over one cardiac cycle (i.e., between two beats of the heart, as represented by points T1 and T2). If we apply the pressure voltage waveform to an integrator, the output of the integrator will be the MAP.

Researchers sometimes want to know the derivative of blood pressure, or dP/dT. Because the waveform is constantly varying, its rate of change, hence its derivative, is also constantly changing. Thus, we can only talk about the derivative at a particular point, such as P1 in Figure 15-3C. We calculate that on paper by taking the slope of a line that is tangent to P at point P1. A differentiator will yield a continually varying voltage that represents the rate of change. Some medical blood pressure monitors are equipped to output voltages representing diastolic pressure, systolic pressure, mean arterial pressure (MAP) and pressure derivative (dP/dT). Later in this chapter we will develop the block diagram of the monitor using some of the circuits discussed.

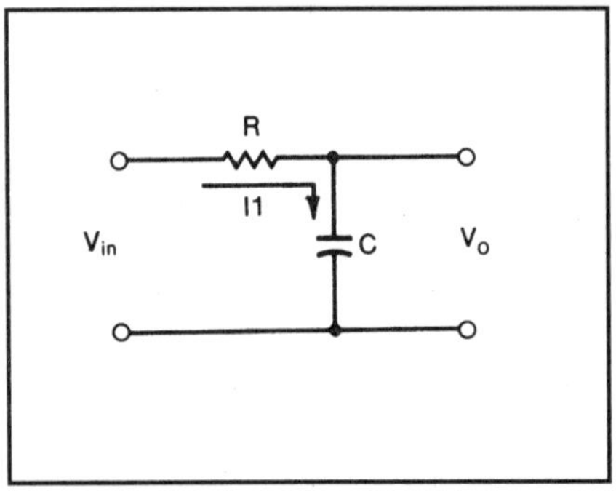

Figure 15-4. R-C integrator circuit.

Electronic Integrator Circuits

Electronic integrators must find the time average of the input waveform. The simplest form of integrator is the resistor capacitor (RC) network shown in Figure 15-4. In this circuit, the output voltage changes as the charge I1 accumulates in the capacitor. The rate of increase in V_o is dependent upon the input voltage and the values of R and C. If you recognize the circuit as a low-pass filter, you are correct. The integrator also functions as an LPF with a roll-off of -6 dB/octave above the -3 dB knee frequency (1/2πRC).

The electronic integrator is shown in Figure 15-5. The active element is an operational amplifier, and it has a capacitor in the feedback network. This circuit is basically the inverting follower of Figure 15-1 with the feedback resistor (R2) replaced by a capacitor. The transfer equation of this circuit is:

$$V_o = \frac{-1}{\omega RC} \int_{t=0}^{t} V_{in} dt \qquad eq.\ (15\text{-}3)$$

— where:

R is the resistance in ohms (Ω).
C is the capacitance in farads (F).
V_{in} is the input signal in volts (V).
dT indicates that we are integrating with respect to time.
ω is the frequency factor 2πF (often deleted when the frequency is close to DC).

The constant factor -1 indicates that the circuit is inverting, so a positive-going input voltage produces a negative-going output voltage — and vice versa.

The gain of this circuit is -1/ωRC. Keep in mind that with small values of R and C, this gain can be very, very high — with resulting problems for the designer. Consider an example.

Example

What is the gain of an inverting integrator in which R = 10K and C = 100 picofarads?

GAIN = -1/RC
GAIN = -1/([10,000 ohms] x [100 x 10^{-12} farads])
GAIN = -1/0.000001
GAIN = -1,000,000

With a gain of 1,000,000 we find that with a small error or DC offset in the input signal, the output will be tremendous. For example, suppose we have a 1 volt peak-to-peak sine wave that has a 10 mV DC offset (not very much). This DC offset will be integrated with a gain of 1,000,000 so the 10 mV becomes (10 mV) x (1,000,000) = 10,000 volts. Of course, the output of the op-amp is limited to about 10 or 12 volts — so the output will saturate really quick. The normal offset of a 741 will saturate the output of such an integrator so fast you'll think it's shorted! In general, the rule of thumb is to make the time-constant of the RC network long (i.e., 5X) relative to the period of the input signal.

Figure 15-6 shows some common electronic signals at the input and output of an electronic integrator. In each case, the upper trace is the input and the lower trace is the output. For the case of the sine wave shown in Figure 15-6A, the output is phase shifted 90 degrees to become a cosine wave. Thus, the output is in quadrature with respect to the input signal.

The integrator action on square waves is shown in Figure 15-6B. Because the amplitude of a square wave is constant, the integrator output will rise at a constant rate until the square wave drops low again. At that time, the slope of the output waveform changes and starts decreasing. Thus, the integrator makes a triangle waveform out of a square wave. Incidentally, this technique is used by function generators to form triangle waveforms.

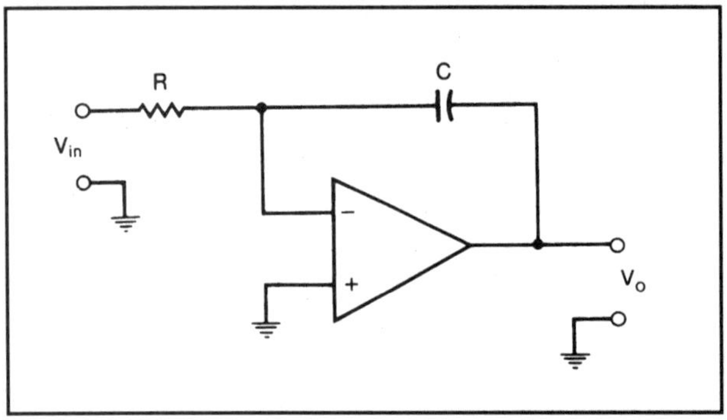

Figure 15-5. Active "Miller" integrator (textbook version).

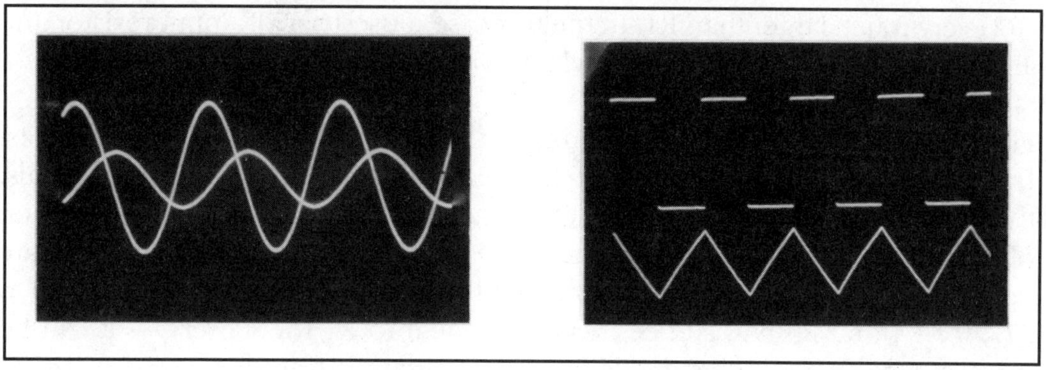

Figure 15-6. Typical integrator response for signals: A) sine wave; B) square wave.

A Practical Integrator

Almost all textbooks on linear integrated circuits or operational amplifiers show the circuit of Figure 15-5. There is only one problem with the circuit — with real operational amplifiers it doesn't work! Unfortunately, some articles and books don't tell you what the problem is, and how to deal with it. When I first started building integrators I discovered the problem the hard way.

The principle problem in practical integrators is that the DC offset voltage normally present at the output of real operational amplifiers will charge capacitor C1, and thereby soon saturates the op amp. The output voltage will start rising immediately after turn on, and will soon be off the scale. If you have a signal applied to V_{in}, then the output will show that signal with a constantly increasing DC offset potential.

One tactic used to cancel the effects of offset is to use an operational amplifier that has a very low offset potential, and no input bias current (or very little). For low-cost applications, the CA-3140 BiMOS op amp (which uses MOSFET input transistors) is a good selection (see Figure 15-7). Devices of the 741 family are almost useless for integrator service.

Another tactic is to connect a resistor across the integrator capacitor in order to keep the DC from building up. This tactic is especially useful for integrators that see periodic input signals. The rule of thumb is to make the shunt resistor

(R2) very much larger than R1. In the test case, I used a 470K input resistor and an 18 megohm shunt resistor — which worked quite nicely.

Finally, we may also have to use an offset null potentiometer in some circuits. In my test case, with 400 hertz sine, square and triangle wave input signals, the potentiometer was not needed. Other cases, however, may require a counter offset provided by R3. Although the values of the resistors in this network are dependent upon the application, most of the time a 5K value for R3, and 10 to 27K for R4 and R5 will suffice. Make R6 equal to R1 for starters — it can be increased or decreased as needed after the circuit is tested.

To adjust the potentiometer, short the input of the integrator (making $V_{in} = 0$). Adjust R3 for a potential of zero volts at point "A." Momentarily close S1 to force the output voltage to zero. If the output voltage rises to either positive or negative values after S1 is opened. Then adjust R3 to cause the rate of increase to slow down to zero. Again close S1 and see if the output voltage changes. Repeat the procedure until the output voltage remains at zero following every closing of S1.

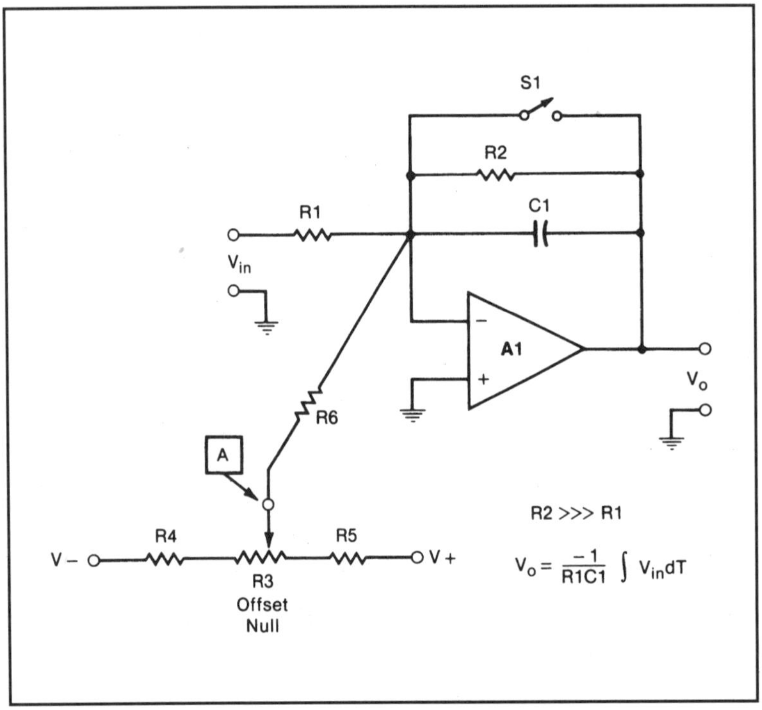

Figure 15-7. Practical active Miller integrator.

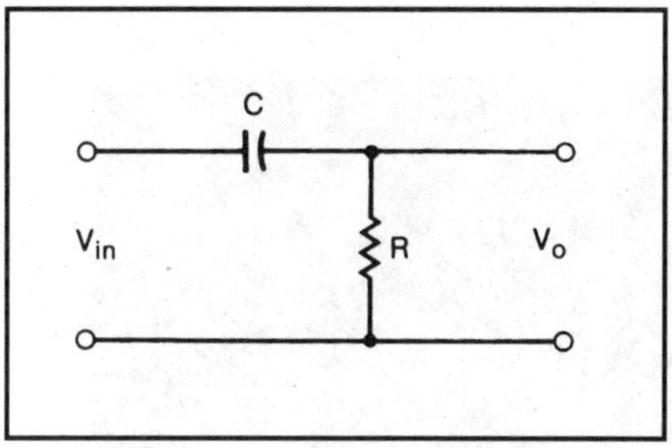

Figure 15-8. R-C differentiator circuit.

In normal operation, switch S1 is used to reset the integrator after it performs an operation. It is used in some instruments where a value is calculated, but is only occasionally needed in cases where an integrator sees a periodic signal with no DC offset component.

Electronic Differentiators

Recall from above that the differentiator is used to produce a signal that is proportional to the rate of change of the input signal. Thus, for a constant DC level, the output of the differentiator is zero. As the rate of change increases, the output potential also increases. For sine waves, the differentiator circuit will output a quadrature signal (in a manner similar to the integrator, but of opposite phase shift). Figure 15-8 shows the basic resistor-capacitor (RC) differentiator circuit.

As long as the input signal period is short compared with the RC time-constant, the output will be the differentiated input signal. You may recognize this circuit as an RC high-pass filter — you're right. The differentiator acts as a high-pass filter with a knee frequency of

$$F_{Hz} = \frac{1{,}000{,}000}{2\pi RC} \qquad \text{eq. (15-4)}$$

— where:

F is in hertz (Hz).
R is in ohms (Ω).
C is in microfarads (μF).

At frequencies above the knee frequency the response of the circuit rolls off at a rate of -6 dB/octave.

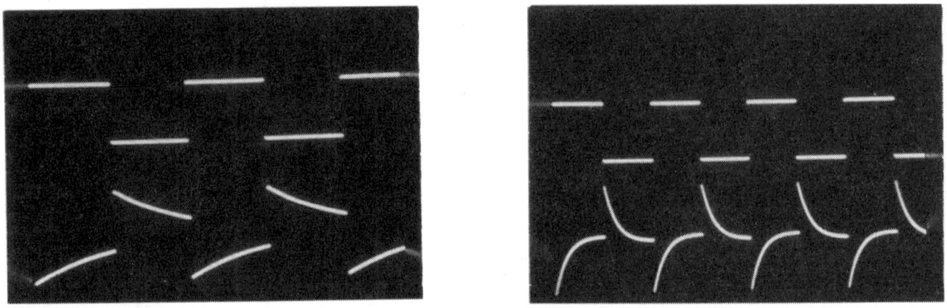

Figure 15-9. Effect of differentiator on signals: A) long time constant; B) short time constant.

Figure 15-9 shows the effects of an RC integrator on square wave signals. The square was selected because its leading and trailing edges have about the fastest rise times producible from our signal generator. In each case (Figure 15-9A and 15-9B), the bottom trace is the square wave input, while the top trace is the differentiated waveform.

In Figure 15-9A we see the case for an RC time-constant that is long compared with the period of the signal. The top trace shows some high frequency roll-off. As the frequency is increased (RC = constant), however, we get the situation shown in Figure 15-9B. Here the period is very short compared with the time-constant. Notice the pronounced effect of the differentiator — the output signals are almost positive and negative spikes. The distortion of the input square wave "flatness" on this trace is a measurement problem with my probe, and was not due to differentiation.

You will see circuits such as Figure 15-8, with a short time constant, in certain applications where the differentiation is sought, but only by implication. For example, in the trigger input of a monostable multivibrator, the "input" to the RC network is a level change caused by a switch. Such a level change resembles a square wave, so is treated something like Figure 15-9B. Thus, the short duration pulse needed to trigger a one-shot is derived from a long duration operation such as pressing a button.

An example of an active differentiator is shown in Figure 15-10A. This operational amplifier circuit is similar to the integrator case, but has the roles of the resistor and capacitor reversed. The output of this circuit (in calculus notation) is:

$$V_o = -RC\frac{dV}{dT} \qquad eq.\ (15\text{-}5)$$

— where:

R is in ohms (Ω).
C is in farads (F)
[Note: The units of the product RC is seconds, s]
V_o and V_{in} are in the same units (i.e., volts, mV, uV).

Like the operational amplifier, the differentiator circuit shown in textbooks is not quite right for real operational amplifiers. Because the differentiator is a high-pass filter, and because of certain instabilities in the circuit, we need to roll-off high frequencies above the point where they are actually needed. Thus, we need capacitor C2 (Figure 15-10B) to roll off high frequencies. We also need a 50 to 250 ohm resistor in series with the input (R2).

Precise Rectifiers

A rectifier is a device that passes current in only one direction. Thus, it can convert bidirectional alternating current (AC) signals to unidirectional pulsating DC (see waveforms in Figure 15-11). Normally, ordinary silicon diodes are used for rectifier service both in power supplies and in signal applications. Although the type of diodes used in both applications are different, the principles are the same.

At low signal levels, however, there is a slight problem. The ordinary diode has a nonlinear "knee" region between 0 volts and a certain junction potential (0.6 to 0.7 volts in silicon diodes). This potential is low in power supply applications, but can be quite significant in low-level signal applications (and, in fact, may exceed the signal voltage).

Figure 15-11 shows a method for serving out the junction potential of the diode by using an operational amplifier. The gain of this inverting follower is

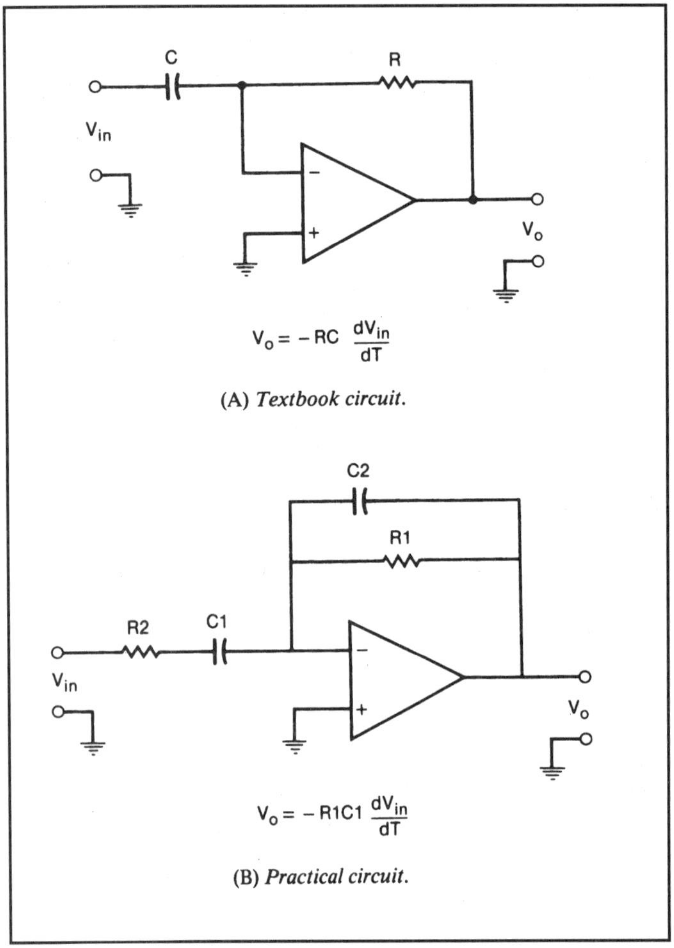

Figure 15-10. Active differentiator: A) textbook example; B) practical circuit.

approximately one-half. Diode D1 is connected so that it will pass only positive output signals from the operational amplifier; negative signals at pin no. 6 reverse-bias the diode, so are blocked. When the input signal is negative, then the output voltage will be positive, with an amplitude of one-half the input signal. However, on positive input signals, the signal at the op amp output is negative, so is blocked. In this case, however, the circuit resistors form a voltage divider with a transfer function of:

- $$V_o = \frac{V_{in}R4}{R1+R2+R4}$$ *eq. (15-6)*

Thus, if (R1 + R2) = R4, the transfer function is 1/2. The output voltage V_o is the same $V_{in}/2$ as seen in the negative input case. With the values shown in Figure

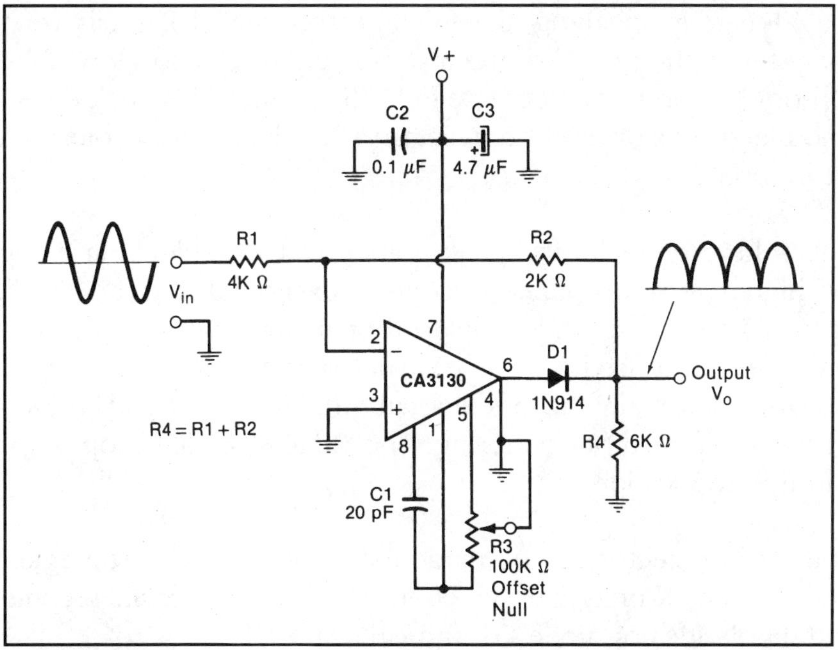

Figure 15-11. Precise rectifier circuit.

15-11, we can expect to process signals up to about 200 kHz. I have successfully built CA-3130 based precise rectifiers for 2 MHz based with smaller resistor values. It should be possible to make a 10.7-MHz AM detector using this method — a project that I am working on for a "home brew" spectrum analyzer.

Peak Detector Circuits

A peak detector is a circuit that will hold the highest amplitude voltage applied to the input. Figure 15-12 shows a simple example of such a circuit. The operational amplifier can be almost any form of high input impedance device.

The holding device is capacitor C1. The charge voltage across this capacitor will be the high potential applied. A low-leakage (i.e., very high reverse resistance) diode (D1) prevents the low output impedance of the op amp from discharging C1.

When the input voltage drops below the previous peak, charge on C1 leaks off through resistor R3, so the capacitor voltage takes on the new "peak" after a period of time. Generally speaking, the RC time-constant of R3C1 is very, very long compared with the period of the input signal. If R3 is too low, then there will be a "droop" error; i.e., a decrease in V_o too soon. Some peak detector circuits use no bleed resistor (R3), but reset the capacitor to zero volts with a switch (see S1 in inset to Figure 15-12).

Three factors affect the usefulness of a peak holder. First, the blocking diode must have a very high reverse resistance. Second, capacitor C1 must be a low-leakage type (for fast signals, use a mica, polycarbonate or polystyrene capacitor for C1). Third, the circuit used to detect V_o (not shown) must have a very high input impedance. Normally, an operational amplifier unity gain noninverting follower is used for output buffering. The BiMOS or BiFET op amps are best suited to this application.

A negative output peak detector can be made using either of two strategies. First, we can always place a unity gain inverter ahead of the circuit. Second, we can reverse the polarities of diode D1 and capacitor C1.

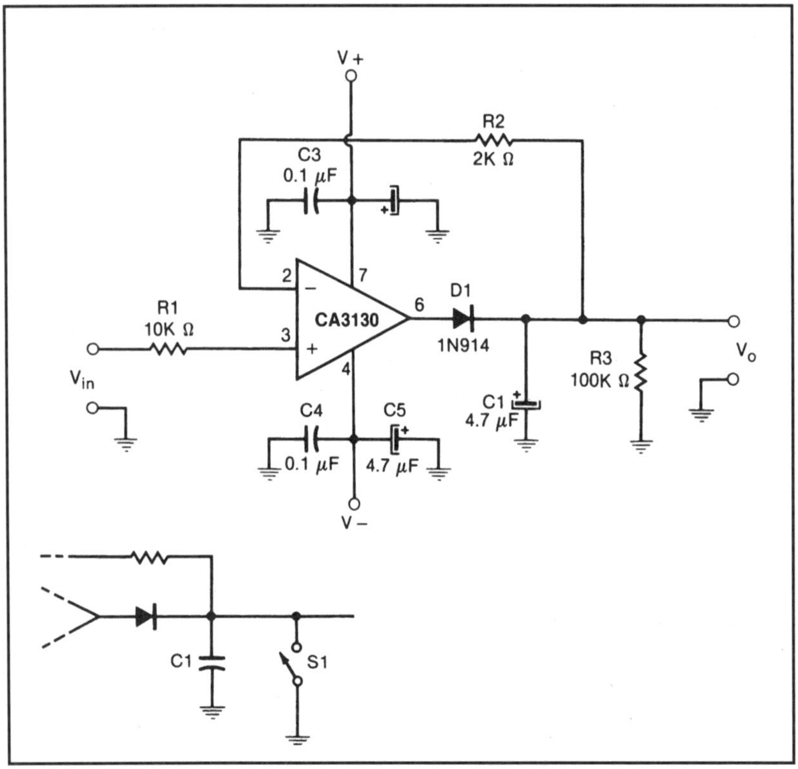

Figure 15-12. Peak detector circuit.

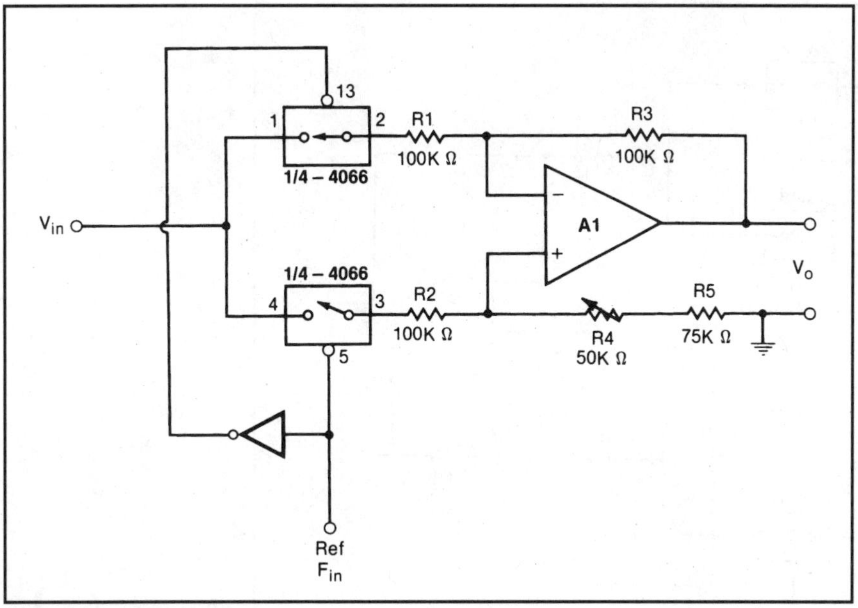

Figure 15-13. Simple quadrature detector.

Quadrature Detector

A quadrature phase detector is used to demodulate AM signals in instrumentation circuits. In a moment, we will consider a practical instrumentation application that uses this type of circuit.

Figure 15-13 shows the phase detector circuit. The active element is a unity gain DC differential amplifier with common mode rejection ratio adjustment (R4). Each input is controlled by a CMOS electronic switch. In early circuits, a vibrator switch was used for this purpose, but today CMOS switches are preferred. The input signal voltage is amplitude-modulated, and has a "carrier" frequency of F_{in}. The reference signal at the same frequency is used to drive the switch. An improved version is shown in Figure 15-14. In this case, four CMOS switches are used for the same purpose. A single CMOS 4066 device can be used for this circuit.

An Instrumentation Example

Now let's consider a practical example of a real instrument that uses some of the circuits discussed in this chapter. Earlier (Figure 15-3C) we used the hu-

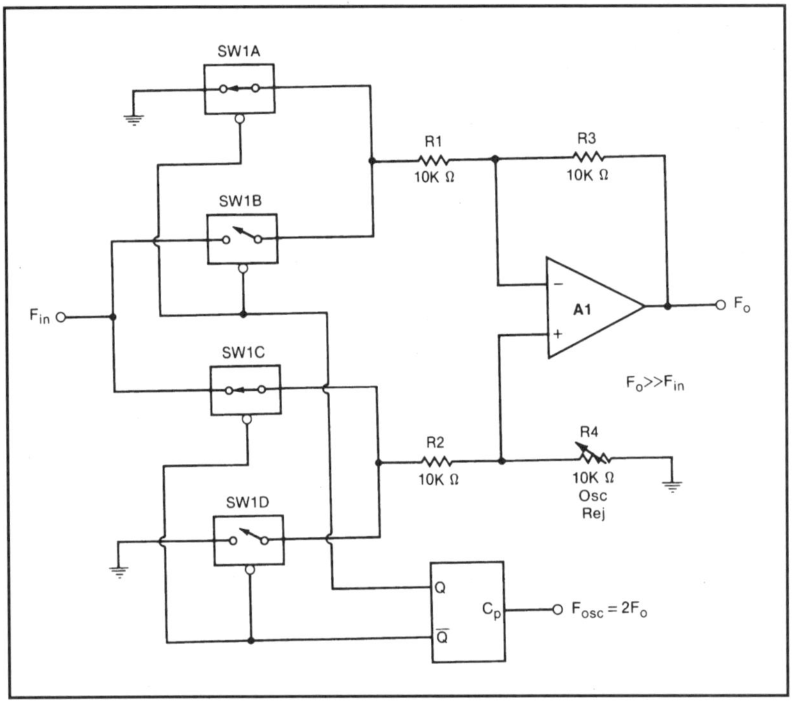

Figure 15-14. Improved quadrature detector.

man blood pressure waveform as an example, so let's make our instrument an arterial blood pressure meter. The curve shown in Figure 15-3C graphs a voltage that represents blood pressure; the waveform is more or less typical of human arterial waveforms. Figure 15-15 shows the block diagram of a blood pressure meter.

The transducer (XDUCER) is either a piezoresistive Wheatstone bridge or a linear variable differential transformer (LVDT). In either case, it is excited by a 1000-Hz reference voltage (V_{ref}). The differential outputs of the transducer are applied to an AC differential amplifier, which builds it up to about 1 volt or so. A DC representation of the waveform (which actually looks like Figure 15-3C) is produced at the output of a quadrature phase detector, and a low-pass filter that removes residual carrier frequency. A DC amplifier is used for scaling of the output voltage, V_o.

Our sample instrument is a super-grade research instrument instead of a mere clinical device, so we need the following signal processing stages: systolic, diastolic, mean arterial pressure (MAP) and dP/dt.

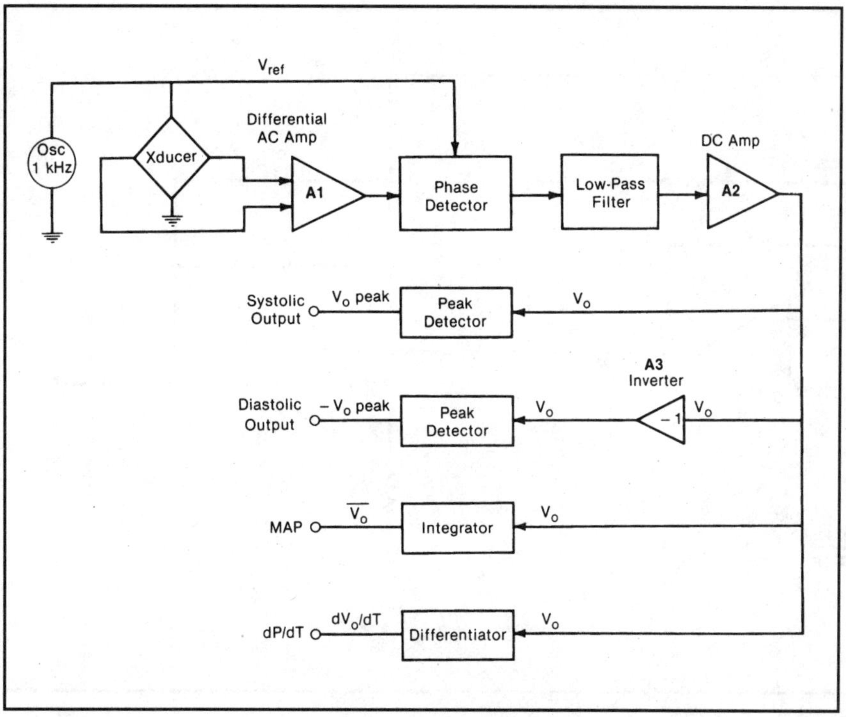

Figure 15-15. Instrumentation example - blood pressure monitor.

The systolic output voltage is a DC level that represents the peak voltage attained by the pressure waveform. Thus, a positive peak detector is used. For diastolic output, on the other hand, we need to invert the waveform and detect the peak that had once been the valley — that is, the diastolic value.

The MAP is merely the time average of the pressure waveform (nurses and doctors deny it — but they learn a simplistic formula that is valid only for perfect waveforms — not for sick people). That waveform is typically derived from a simple op amp integrator in medical instruments. Similarly, the dP/dt signal (the pressure waveform derivative) is derived from an op amp differentiator.

All of the signals except the dP/dt are DC levels proportional to the value being measured. For these, any analog or digital DC voltage reading instrument can be used. For the dP/dt, however, only an oscilloscope or oscillographic paper recorder is permitted.

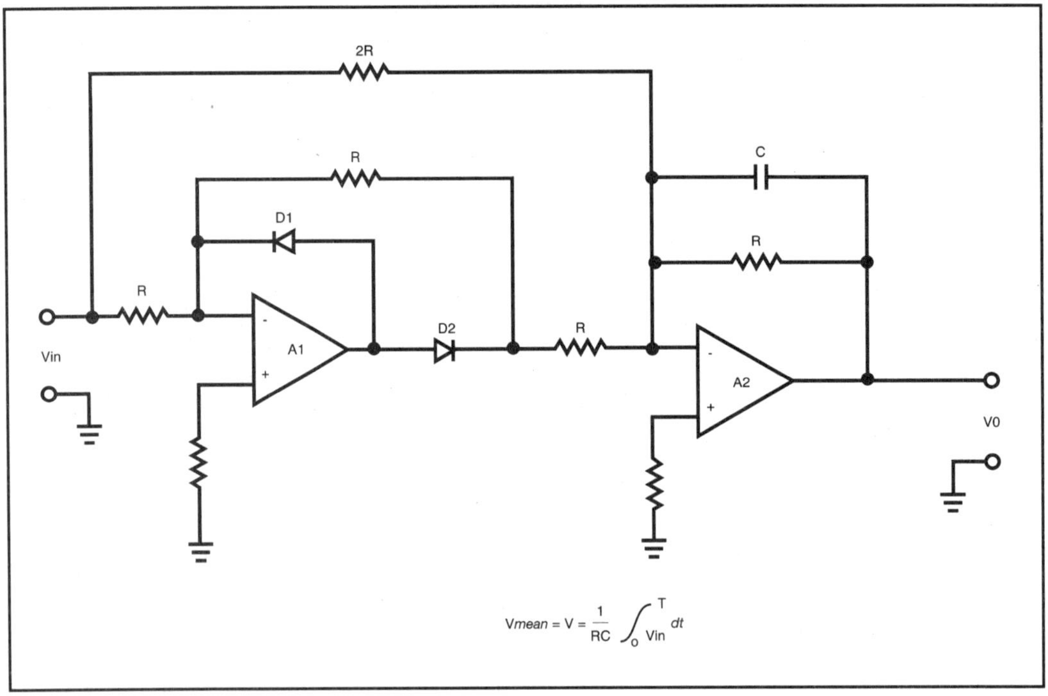

Figure 15-16. Mean value amplifier circuit.

Mean Value Detector

A mean value detector, or time averager, circuit is shown in Figure 15-16. This circuit produces an output voltage, V_o, that is proportional to the mean value of the input signal, V_{in}, over time (the time period is set by the values of R and C). For sine waves, the output voltage approximates the root mean square (RMS) value of the input voltage.

The averaging function is performed by op amp A2, which is connected as a Miller integrator. The integrator is driven by a precise rectifier circuit. One of the advantages of using the precise rectifier, rather than a simple diode, is that the detection is linear rather than square law (the "knee" in the diode curve is feedbacked out). The output voltage, V_{mean}, is given by:

$$V_{mean} = \overline{V} = \frac{1}{RC} \int_{t=0}^{T} V_{in} dt \qquad \text{eq. (15-7)}$$

This circuit is used in precise AC voltage measurements, and in VLF radio studies of the sun. The VLF studies use a 20 to 70 kHz receiver to monitor distant VLF stations. When a solar flare occurs, which results in a sudden ionospheric disturbance (SID) event, the shortwave bands (3 to 30 MHz) die out, but the VLF bands are enhanced. If the operational amplifier used for A1 will handle the frequencies involved, then this circuit produces a DC output voltage proportional to signal strength. Whatever the application, a good operational amplifier, with low output offset voltage, such as the CA-3140, is needed for A2 (and would be a good idea for A1 as well). A good compromise is to use the CA-3240, which contains a pair of CA-3140-class devices. The diodes are 1N914 or 1N4148 silicon devices.

Minimum Value Detector

The minimum value detector shown in Figure 15-17 is used to measure low amplitude signals in the presence of high amplitude transients. I ran across this circuit in dealing with VLF radio receivers. In the solar studies mentioned above, it is normally necessary to measure signal amplitude levels. In some cases, however, there is a lot of lightning static in the air. In fact, in some locations, during certain times of the year, the static might be so severe as to

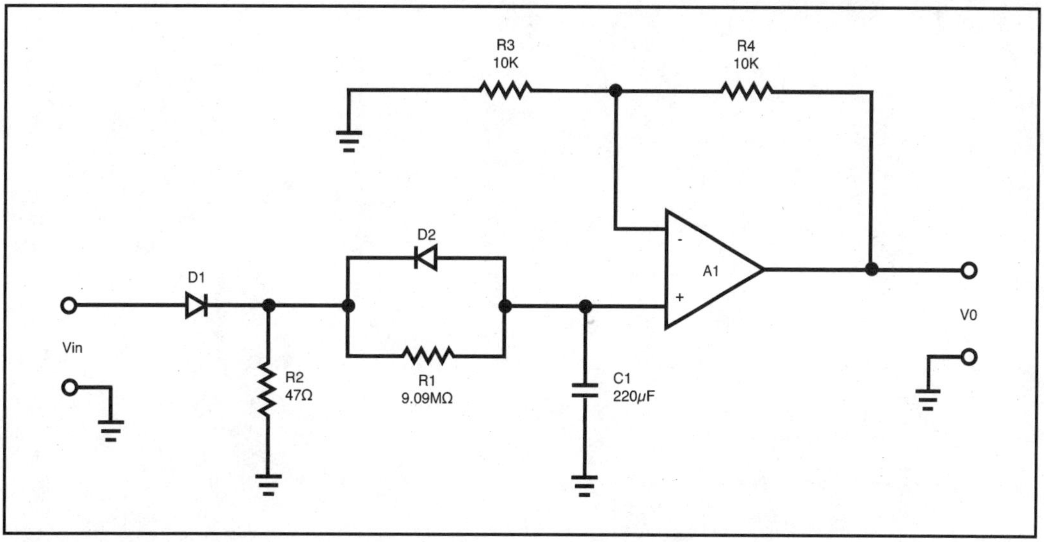

Figure 15-17. Minimum value circuit.

eliminate all experiments on VLF radio bands. However, with this circuit, the static can be eliminated.

The circuit works by having a rectifier drive an integrator, but arranged such that there are two time-constants involved, T1 and T2. Time constant T1 is the product R1C1, and is of the order of 20 seconds; time constant T2 is R2C1, and is of the order 0.01 seconds. With this circuit, fast rise time lightning impulses received from the antenna are snubbed out, while the continuous VLF signal and hiss signals are passed.

Chapter 16

DC Power Supplies for Op Amp Circuits

The DC power supply is critical to the success of any electronic circuit, perhaps especially op amp circuits. This chapter deals mostly with the control and distribution of DC power, with some special emphasis on reference supplies and excitation power supplies needed for linear IC circuits. We will look at the standard DC power supply circuit only in passing for those who may not be familiar with it.

DC Power Supplies

The purpose of the DC power supply is to convert alternating current (AC) to direct current (DC) for use in electronic circuits. Figure 16-1 shows the basic circuit diagram for the DC power supply. Transformer T1 scales the voltage from 120 or 240 VAC to the voltage range needed for the electronic circuits (typically 5 to 28 VDC).

The AC from the transformer is converted to DC in the bridge rectifier BR1. These four-terminal devices take the AC voltage at the terminals marked "AC" and output them through the DC terminals ("+" and "-"). The output of the rectifier is pulsating DC with a peak voltage about 0.9 times the peak AC

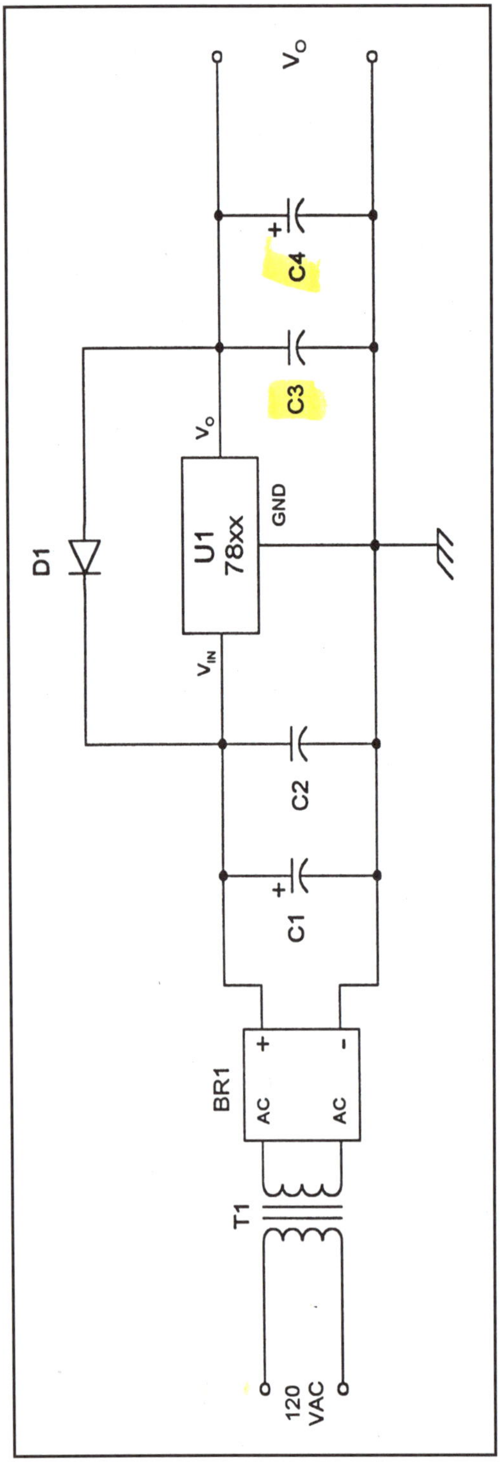

Figure 16-1. Typical regulated voltage supply.

voltage. The voltage rating for the transformer is in RMS voltage, so the output of the rectifier is 0.9 x 1.414 x V_{RMS}, or $1.27V_{RMS}$, where V_{RMS} is the voltage appearing across the secondary winding of the transformer.

The voltage regulator (U1) keeps the output voltage constant despite variations of input voltage or variations of load current. There is usually a 2.5 to 3 volt drop between the V_{IN} and V_O terminals of the voltage regulator. Thus, the transformer secondary RMS voltage must be $\geq (V_O + 3)/1.27$.

The voltage regulator is usually a three-terminal IC device of either LM-340n-xx or 78xx types (essentially the same thing under different nomenclatures). In LM-340 devices, the "n" is replaced with the package style indicator: L for TO-92, K for TO-3, and T for TO-220. The "xx" is the voltage rating. Thus, an LM-340T-5 is a +5 volt regulator in a plastic TO-220 package. For negative output regulators, the LM-320n-xx series are used. In 78xx devices, the "xx" is the voltage rating, so a +5 volt type is 7805, while a +12 volt is 7812. If the device is rated at

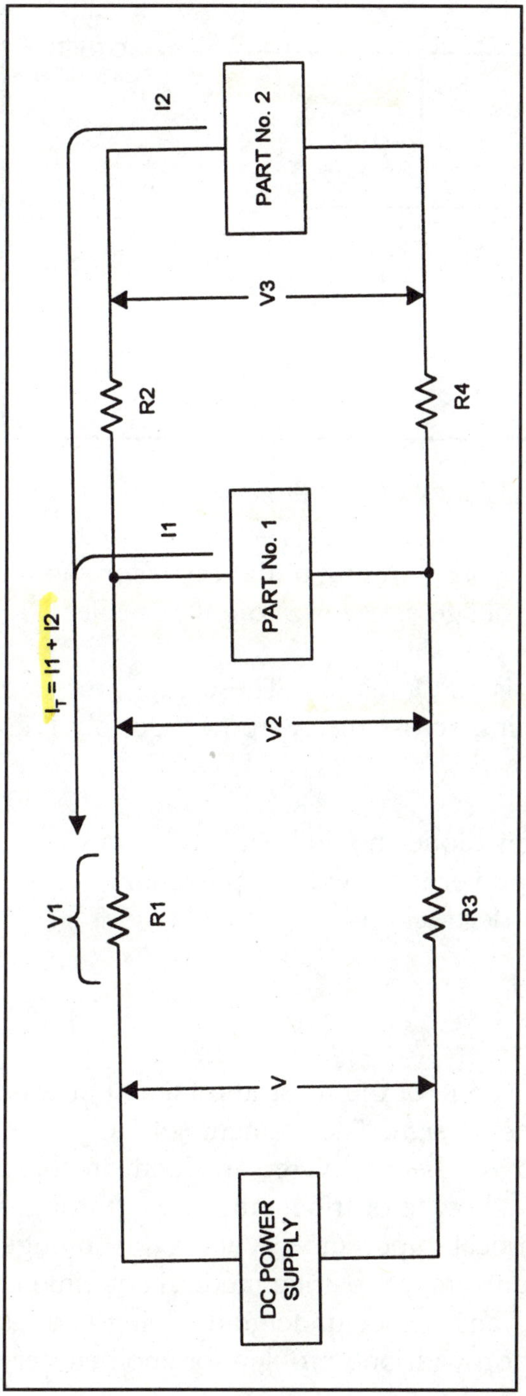

Figure 16-2. The problem when two or more circuits share a DC power supply.

100 mA or so, then an "L" is placed between the 78 and xx: 78L05 is a 100 mA, +5 VDC device. Negative output equivalents are designated 79xx.

The current ratings for most regulators are:

TO-92 100 mA
TO-220 750 mA (1,000 mA with heat sink)
TO-3 1,000 mA (some are 3,000 mA)

Most common devices are either TO-92 (e.g., 78L05), or TO-220 (e.g., 7805), producing 100 mA and 750 mA respectively.

Capacitor C1 in Figure 16-1 is rated at 2,000 µF per ampere, except for low-current devices in which case 1,000 µF is used; for the standard 750-1000 mA supply a 2,000-µF unit is used. The exact rating is not important, so long as the right range is achieved. The output capacitor (C4) is optional, and is used to smooth variations occurring before the regulator can catch up; i.e., transients. This capacitor is rated nominally at 100

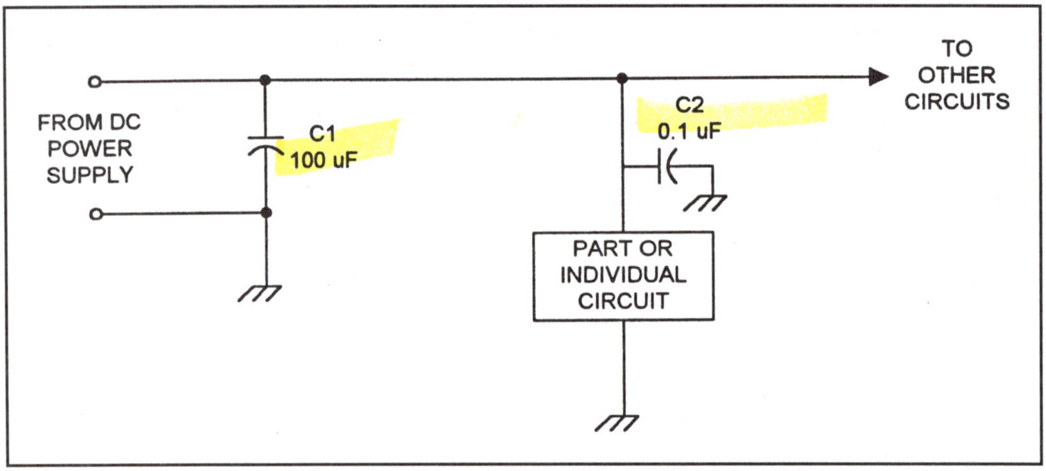

Figure 16-3. A decoupling capacitor (C1) at each stage will work wonders.

µF per ampere. In both C1 and C4, the voltage rating of the capacitor should be at least 150 percent of the applied voltage rating — more if possible.

Capacitors C2 and C3 are used for noise suppression. These capacitors are needed to take care of transients coming across the AC power supply line. Make these capacitors 0.1 µF to 1 µF.

Diode D1 is used only if C4 is used. This diode should be a 1N4007 device. It is used to rapidly discharge C4 when power is turned off, preventing it from biasing the substrate of U1, causing its destruction.

DC Power Distribution

The direct current (DC) power supply is one of the most important, but least considered, aspects of electronic project design. The elements of DC power supply design are well known, and have been covered in depth in these pages. However, what is often missing is how to distribute the power throughout a project. That little matter is of critical importance. When I was in engineering grad school, I worked for the university repairing medical equipment. A fairly large number of grad students and senior undergrads doing various projects found their way to my workshop with one problem or another. Very

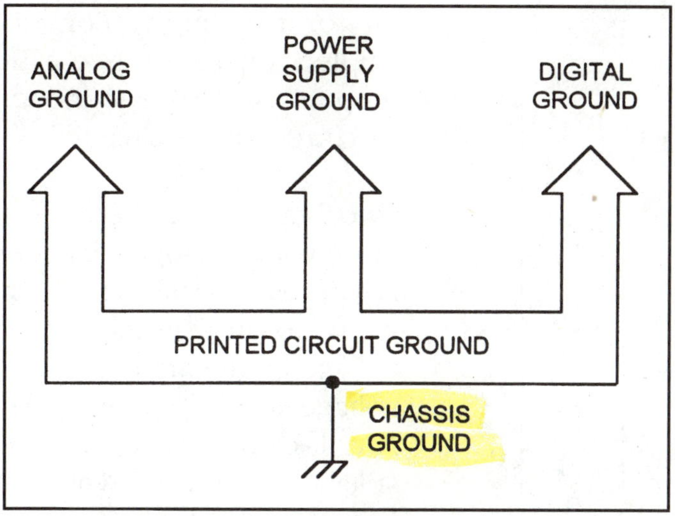

Figure 16-4. Single-point grounding is well-suited to keeping the problems in check.

frequently, the design of the circuit was fine, but the implementation was a real problem. In most cases, they created their own bad situation by paying too little attention to the power supply distribution. The main errors were too small conductors and no decoupling capacitors. One fellow came into my shop with a perf-board covered with about thirty TTL chips, and complained that it worked only some of the time. After ascertaining that the problem was probably not a loose soldering connection or broken wire, I noticed that the wire to each chip was very small (#30 wire-wrap wire), and that there were no bypass capacitors anywhere on the perf-board. I instructed him to put 470 µF at the +5 VDC input line, plus 0.1 µF at each TTL chip — and increase the gauge of the wire distributing power. When he made those changes, the project worked properly.

In the sections below you will find out why these are important.

Figure 16-2 shows a generic electronic circuit DC power distribution system. A DC power supply, which may or may not be voltage-regulated depending on the application, feeds DC to the individual circuits via a common DC power line and a ground. In an ideal world, that would be the end of our concerns. However, in real circuits, there is a DC resistance associated with all of the power line conductors and any connectors used in the system. These resistances are shown as R1 through R4 in Figure 16-2. As current drawn by each part flows through the conductors, a voltage drop will appear across the con-

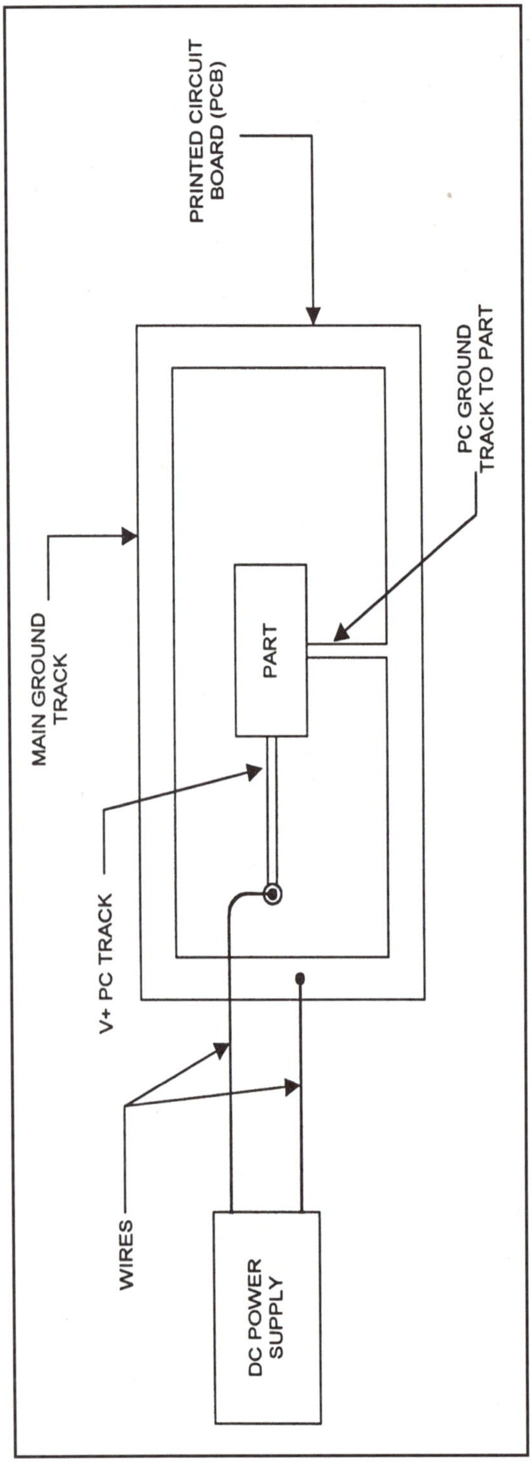

Figure 16-5. Simple DC distribution system.

ductor resistance. For example, voltage V1 is the IR drop across R1 due to the total current in the circuit.

Two things happen because of those voltage drops. First, the actual voltage available to each part is lower than the supply voltage by the amount of the voltage drops. For example, Part-1 can see a maximum voltage (V2) of V-V1. When selecting a value for V you must either take into consideration the voltage drops that will occur, or reduce the resistance to a negligible value by making the conductors large enough.

The second thing that can happen is that varying currents in one part will cause the supply voltage to vary at that part and at other parts. For example, if the DC load represented by current I1 varies with the signal in Part-1, then the total current varies, and that means the voltage drop across R1 also varies. That varying DC supply voltage is applied to Part-2 as its supply voltage. Signal can thus be coupled unintentionally from one stage to another through the DC power

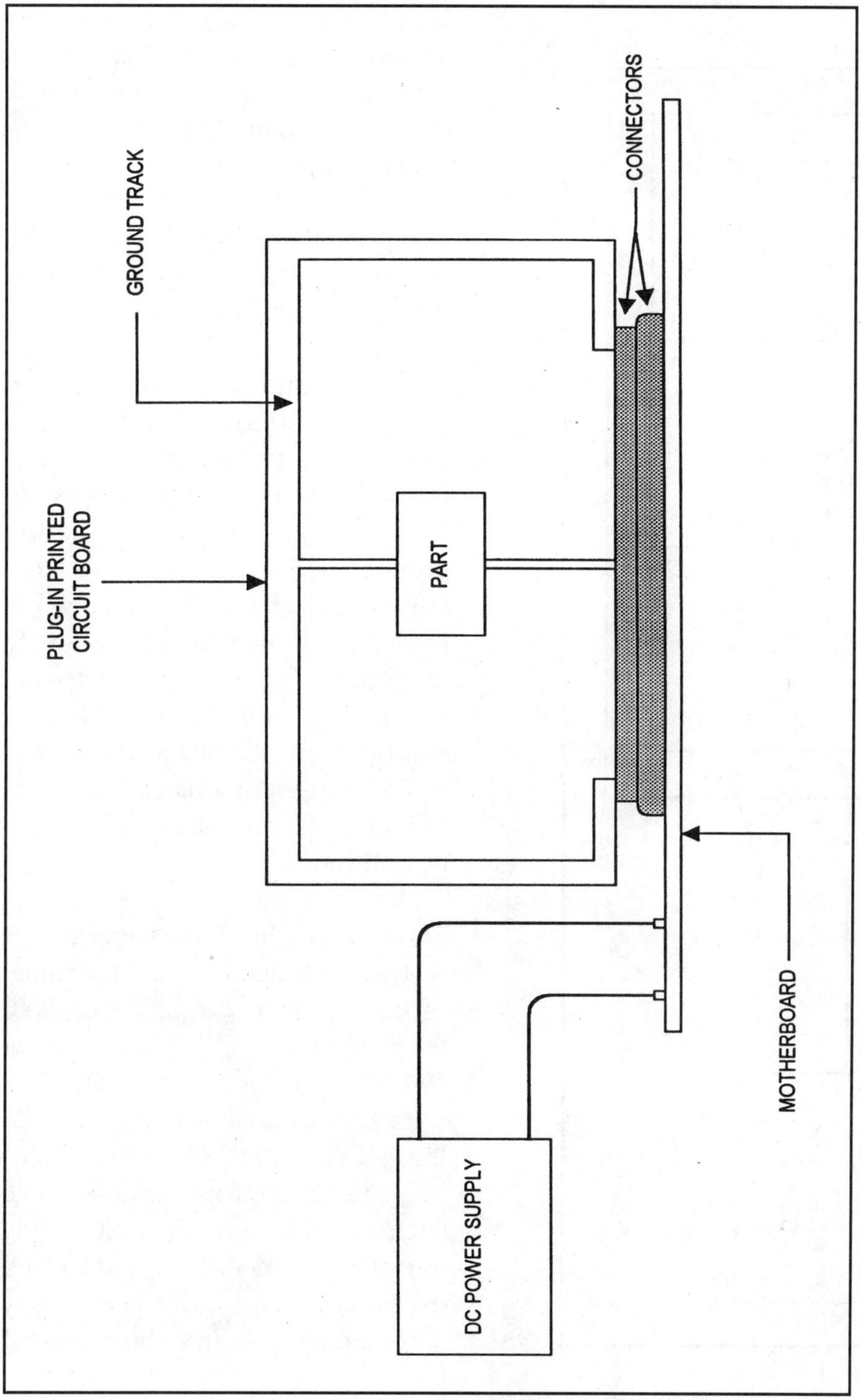

Figure 16-6. Use of a motherboard and plug-in PCBs.

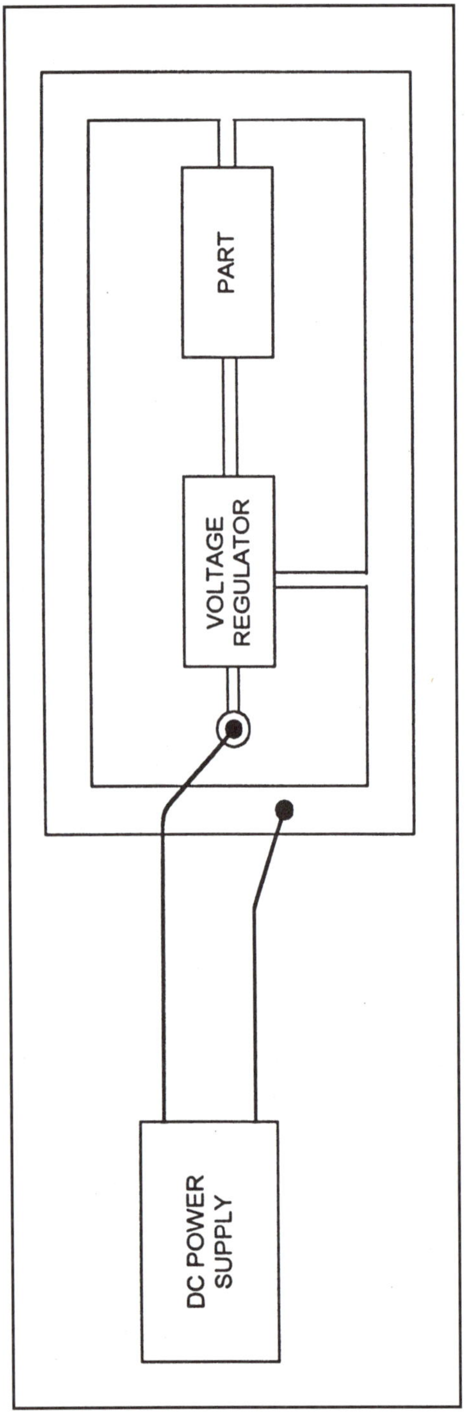

Figure 16-7. The use of an on-board voltage regulator.

supply lines. This effect is usually called power supply sensitivity (PSS), and certain components (e.g., operational amplifiers) have a PSS specification.

The power supply DC return lines, which are usually also the grounds for the circuit, have the same kind of resistances as the “hot” line. If the same conductors are also used as signal grounds (often the case), then variations in ground voltage drops are seen as valid signals between stages; these are called ground loops. The equivalent circuit would show the varying DC from the ground loop in series with the normal signal applied to the stage. In a multistage circuit, the combination of voltage drops and spurious signal paths can be quite complex — and very difficult to troubleshoot.

Fortunately, there are some things that you can do about it. One thing, of course, is to be generous with the wire size used in the circuit for DC power lines and grounds. I've seen problems caused by using #24 hook-up wire where #18 would've been more appropriate. Also, I've seen cases where printed wiring tracks were too narrow and caused excessive voltage drops in relatively short lengths. My first "home brew"

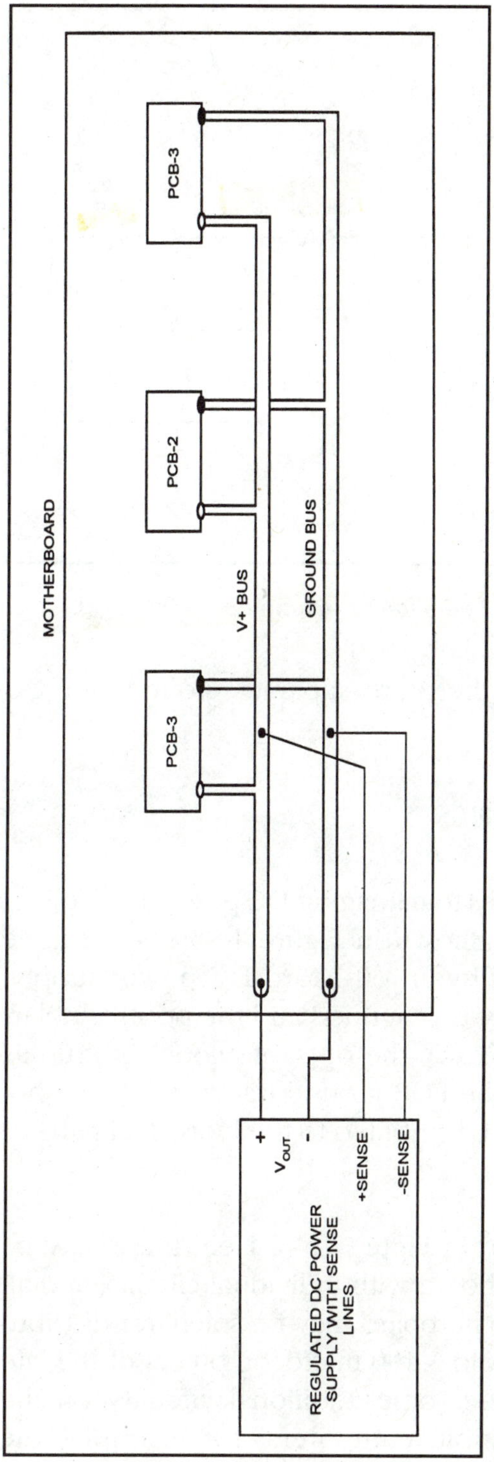

Figure 16-8. Use of DC power supply with sense lines.

microcomputer had that problem. Because it contained a large number of old-fashioned TTL devices, each drawing 25 mA, the current requirement was huge.

Another thing that you can do about the problem is to be generous with the number of capacitors used in the circuit. The capacitors (e.g., C1 and C2 in Figure 16-3) serve two functions: they decouple noise signals on the power line (prevent them from being coupled from stage to stage), and they serve as local reservoirs of current to fill in for short-duration current needs of the circuit being served. It has long been the practice in digital circuits to put a 0.1 μF capacitor across the power terminals of each chip on the circuit board. Putting a 1 μF or higher capacitor at each analog circuit also works well.

Another thing that can be done is to use "single-point grounding" techniques. Each circuit should have its own ground system, and all of the individual ground systems are connected to the chassis at only one point. Also, if you mix different kinds of grounds in the same project, then each type should have its own ground bus.

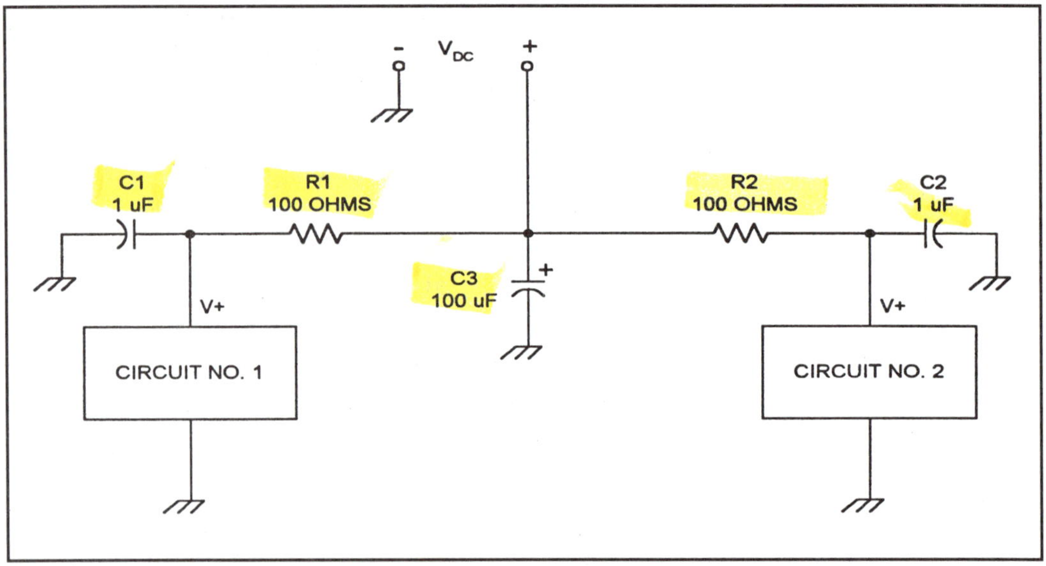

Figure 16-9. Isolation resistors or RF chokes will often improve isolation between stages.

In the example shown in Figure 16-4, there are separate grounds for DC power, analog signals and digital signals.

Power Distribution Schemes for Circuit Boards

There are a number of different approaches to managing DC power on projects with printed circuit boards. The version shown in Figure 16-5 is the simple case of an individual circuit card served by an off-board DC power supply. Although in this case the negative polarity is grounded and the positive polarity is the "hot" line (the most common case), the opposite could be true as well. The common line is connected to the PCB ground tracks, while the hot line is connected to the V+ line. The DC is distributed to the individual parts or sub-circuits through printed circuit tracks.

A different, but related scheme is shown in Figure 16-6. Here there may be more than one individual printed circuit board with individual circuits, and all are plugged into a motherboard. The motherboard may be solely for distribution of signals and DC power (as in the old S-100 microcomputers of the late 1970s and early 1980s), or there may be some additional circuitry on the motherboard. On the IBM PC compatible computers, for example, the motherboard contains the processor and main functions of the board, and the plug-in PCBs perform I/O, video and other functions. In either case, the DC power is distributed to the various printed circuit boards via the motherboard and the plug-in connectors that mate the motherboard to the plug-in boards.

The situation in Figure 16-6 assumes that a single DC power supply serves the needs of all the PCBs plugged into the motherboard. In some cases, however, if current drains are high or there is a lot of digital (or other) noise generated on the PCBs, then one might wish to use an approach like Figure 16-7. This method is called distributed voltage regulation because each plug-in card has its own voltage regulator. The effect of the multiple regulators is to isolate the circuits on the board from ill effects caused by the operation of the other boards. The main DC power supply may be either regulated or unregulated. If a regulated main power supply is used, it is called preregulation, and is capable of a very good degree of power supply ripple reduction.

The power supply sense method is shown in Figure 16-8. A DC power supply with sense lines samples the voltage present at a remote site, and then adjusts its own voltage set-point to put the power supply output voltage high enough to meet the requirement at the remote site. For example, in a microcomputer that uses a motherboard and plug-in cards, it might be true that the processor card (PCB-2) is the most critical of voltage variation, and the memory cards are the least critical (PCB-1 or PCB-3). As a result, the builder will place the sense lines closest to the processor card, causing the DC power supply voltage to be +5 VDC at that point. Closer to PCB-1 it will be higher, and towards PCB-3 it will be lower.

Isolating Circuits on the PCB

Figure 16-9 shows a way to improve the isolation of two or more circuits in a group. In most circuits, one would normally connect all the V+ lines to the positive DC power supply line, resulting in a potential feedback situation due to poor decoupling between stages. However, the isolation between the two circuits can be improved if the V+ line to each circuit has a 10 to 100 ohm resistor (R1 and R2), and two additional decoupling bypass capacitors (C1 and C2). The main V+ line from the DC power supply is decoupled through a higher value capacitor (C3). The exact value of the resistances depends on the current drain, the V+ values required by each circuit and the value of the supply voltage (VDC). At the normal current drains for each circuit the voltage dropped across the isolation resistors should be only a small percentage of the total voltage. In radio frequency (RF) circuits the resistors may be replaced with radio frequency chokes (RFCs).

Index

SYMBOLS

100 HZ INVERTING AMPLIFIER 47
1N400X-SERIES DIODES 21
3140 BIMOS OP AMP 15
3652 123
4066 DEVICE 118
600 OHM AUDIO CIRCUITS 165
600-OHM LINE 148
722 126
741 69, 89, 93, 108, 152
7805 245
78L05 245
78XX 244
8-PIN METAL CAN 44
8-PIN MINIDIP PACKAGE 44

A

A-703 4
AC 107, 243
AC AMPLIFIER 47, 77, 103, 104, 107
AC COUPLED IA 112
AC DIFFERENTIAL 238
AC IA CIRCUITS 105
AC INPUT SIGNAL 139
AC INSTRUMENTATION AMPLIFIERS 103, 105
AC LINES 116
AC MIXER CIRCUIT 148
AC POWER LINES 117, 118
AC POWER MAINS 113, 115, 121, 123
AC POWER SUPPLY LINE 246
AC SCALE 108
AC SIGNALS 12, 59, 75, 92, 135
AC SUMMER CIRCUIT 148
AC VOLTAGE MEASUREMENTS 241
AC-COUPLED AMPLIFIER 41
AC-COUPLED AMPLIFIER CIRCUIT 57
AC-COUPLED INSTRUMENTATION AMPLIFIER 107
AC-COUPLED INVERTING FOLLOWERS 41
AC-COUPLED NONINVERTING AMPLIFIER 55
AC-GND-DC COUPLING 106
ACTIVE DEVICE 198
ACTIVE DIFFERENTIATOR 233
ACTIVE FILTER CIRCUIT 192
ACTIVE FILTERS 185
ADAPTATIONS 105
AF INPUT SIGNAL VOLTAGE 166
AIR VARIABLE 189
ALTERNATING CURRENT (AC) 233, 243
ALTERNATIVE CIRCUITS 21

ALUMINUM 20, 23, 90
AMATEUR RADIO OPERATORS 187
AMPERES 7, 129
AMPLIFICATION 30
AMPLIFIER 5, 8, 9, 15, 25, 26, 27, 28, 30, 31, 47, 49, 50, 55, 60, 62, 64, 70, 76, 97, 107, 108, 113, 115, 118, 122, 137, 144, 157, 159, 165, 169, 175, 177, 198, 201, 202
AMPLIFIER ARRAY 164
AMPLIFIER CIRCUITS 73, 99, 144
AMPLIFIER FUNDAMENTALS 5
AMPLIFIER INPUT 43
AMPLIFIER INPUT IMPEDANCE 52
AMPLITUDE 118, 214
AMPLITUDE SIGNALS 241
AMPLITUDE-MODULATED 118
ANALOG 183, 239
ANALOG CIRCUIT 251
ANALOG FILTERS 190
ANALOG INPUT SIGNAL 122
ANALOG INTEGRATED CIRCUIT 203
ANALOG METER 108
ANALOG NOTCH FILTER 187
ANALOG OUTPUT 51
ANALOG SIGNAL 118, 252
ANALOG SIGNALS 252
ANALOG WAVEFORMS 214
ANALYSIS METHOD 25
ANTENNA 242
ANTENNA INPUT TERMINALS 30
APPLICATION CIRCUIT 125
APPLICATIONS 95, 150, 201, 233
APPLIED VOLTAGE 177
ARTERIAL BLOOD PRESSURE METER 238
ARTERIAL BLOOD PRESSURE MONITORS 226
ARTERIAL WAVEFORMS 238
ARTIFACT 107
ASTABLE MULTIVIBRATORS 197, 217
ATMOSPHERIC NOISE 30
ATTENUATION 50
ATTENUATOR NETWORK 211
AUDIO 9, 152, 165
AUDIO AMPLIFIER CIRCUITS 151
AUDIO AMPLIFIERS 7, 92, 151, 160, 167
AUDIO APPLICATIONS 16
AUDIO CIRCUITS 154
AUDIO COMPRESSION AMPLIFIER 169
AUDIO CONSOLE 162
AUDIO DEVICES 16
AUDIO MIXER 42, 148, 162, 164
AUDIO MIXER CIRCUIT 162
AUDIO PREAMPLIFIERS 151, 156
AUDIO SIGNAL 170
AUDIO SPECTRUM 151, 154
AUDIO TAPER POTENTIOMETER 163
AUDIO VOLTAGE AMPLIFIERS 170
AUDIO VOLTMETER 92
AUTOMATIC GAIN CONTROLLED (AGC) 181
AUTOMOTIVE EQUIPMENT 135
AVERAGE VALUE GENERATION 95

B

BALANCES 88
BANDPASS CIRCUIT 187
BANDPASS FILTER (BPF) 187
BANDPASS FILTER FUNCTION 193
BANDPASS OUTPUT 192
BANDWIDTH (BW) 26, 31, 44, 75, 93, 156, 187
BANDWIDTH FREQUENCY RESPONSE 76
BASE BIAS CURRENTS 28, 87
BASE CURRENT 6
BASIC CONFIGURATION 35
BASIC DESIGN PROCEDURE 43
BASS 159, 160
BASS BOOST 159
BATTERY OPERATED EQUIPMENT 19
BATTERY-POWERED IA 117
BAXANDALL CIRCUIT 160, 161
BAXANDALL TONE CONTROL 160
BIAS 28, 138
BIAS CURRENT 44, 130, 131, 132, 149, 150, 229
BIAS VOLTAGE 12
BIDIRECTIONAL ALTERNATING CURRENT (AC) 233
BIFET 44, 47, 69, 89, 90, 99, 100, 236

BIFET OP AMPS 28, 87
BIMOS 44, 47, 69, 89,
90, 99, 100, 107, 161, 222, 236
BIMOS DEVICES 28
BIMOS OP AMPS 87
BINARY INPUT 216
BINARY PATTERNS 217
BIPOLAR DC POWER SUPPLY
11, 123, 150
BIPOLAR INPUTS 28
BIPOLAR NPN TRANSISTORS 87
BIPOLAR PNP TRANSISTORS 87, 118
BIPOLAR POWER SUPPLY 11
BLOCKING DIODE 236
BLOOD PRESSURE 226, 238
BLOOD PRESSURE METER 238
BLOOD PRESSURE WAVEFORM 226, 238
BOOSTING 50
BREADBOARD 77
BREADBOARDING 19
BRIDGE RECTIFIER 243
BRIDGED-TEE OSCILLATORS 205
BROADCAST APPLICATIONS 165
BROADCASTING 9
BUF-634 180
BUFFER AMPLIFIER 50, 180
BUFFERING 50, 51, 54, 95
BURR-BROWN 180
BURR-BROWN 3652 123
BURR-BROWN CORPORATION 9
BURR-BROWN INA-102 110
BURR-BROWN INA-105 95
BURR-BROWN ISO122P 125
BYPASS CAPACITORS 223

C

CA-3048 164
CA-3048 AMPLIFIER 208
CA-3080 130
CA-3080/CA-3080A 130
CA-3130 180, 235
CA-3140 28, 107, 161, 180, 195, 241
CA-3140 DEVICE 30
CA-3160 180
CA-3240 107, 195, 241
CA-3600E 175
CABLE 51
CAPACITANCE 40, 77
CAPACITIVE REACTANCE 75, 81, 90, 146
CAPACITOR 12, 39, 48, 57, 59, 60,
75, 104, 108, 139, 141, 146,
149, 154, 159, 189, 192, 209,
214, 233, 235, 236, 245, 253
CAPACITOR INPUT 57
CAPACITOR VALUE 149
CAPACITOR VOLTAGE 235
CAPACITORS 19, 20, 61, 104, 152, 178,
179, 181, 183, 185, 189, 206, 246,
251
CAR 135
CARDIAC CYCLE 226
CARDIAC MONITORS 115
CARRIER AMPLIFIER 118, 120
CARRIER METHOD 121
CARRIER OSCILLATOR 118
CARRIER SIGNAL 118, 119, 120
CARRIER TECHNIQUE 117
CARRIER-OPERATED IA 117
CASCADE STAGES 78
CASSETTE TAPE PLAYERS 154
CASSETTE TAPE PREAMPLIFIER 76
CATHODE RAY OSCILLOSCOPE 106
CDA 17, 135, 136, 138, 141, 144, 150
CDA 600-OHM LINE DRIVER AMPLIFIER
148
CDA CIRCUITS 139
CDA CONFIGURATION 136
CDA DIFFERENTIAL AMPLIFIER 146
CDA MIXER 148
CERAMIC 23
CHANNEL 148, 162, 165
CHARTS 80
CHASSIS 116, 135
CHASSIS GROUND 125
CHIP 21, 159, 213, 251
CHIP BODY 183
CIRCUIT 6, 20, 25, 28, 31, 36, 41,
47, 50, 52, 59, 77, 87, 89, 91, 92,
98, 99, 100, 117, 118, 125,
133, 136, 143, 145, 147, 148,

150, 152, 154, 157, 159, 160, 166, 170, 172, 181, 187, 192, 198, 203, 209, 221, 230, 233, 235, 248, 251, 253
CIRCUIT BOARDS 252
CIRCUIT CONNECTIONS 172
CIRCUIT ISOLATION 253
CIRCUIT OPERATION 26
CIRCUIT REQUIREMENT 54
CIRCUIT RESISTORS 234
CIRCUIT SYMBOLS 9, 116, 135
CIRCUIT WORKS 242
CIRCUITRY 51
CIRCUITS 9, 32, 42, 45, 73, 77, 139, 150, 182, 201, 208, 221, 254
CLAIREX 169
CLASS-A 5
CLASS-AB 6
CLASS-B 5
CLASS-C 6
CLASSES 14
CLINICAL DEVICE 238
CMOS 209, 237
CMOS ELECTRONIC SWITCHES 118
CMOS SWITCHES 237
CMRR 84, 88, 90, 93, 102, 109
CMRR ADJ CONTROL 105
CMRR ADJUSTMENT 90
COLLECTOR 86
COLLECTOR CURRENT 122
COLLECTOR LOAD RESISTORS 86
COLLECTOR VOLTAGE 86
COLLECTOR-EMITTER CURRENT 136
COLOR DOT 179
COMMERCIAL 123
COMMERCIAL IC IA 110
COMMON EMITTER CONFIGURATION 136
COMMON MODE 237
COMMON MODE REJECTION RATIO (CMRR) 84, 88, 90, 102
COMMON MODE SIGNALS 83, 84
COMMON OP AMP 50
COMMON-CATHODE MODE 202
COMMON-EMITTER BIPOLAR TRANSISTOR 202
COMMON-MODE REJECTION 97
COMMUNICATIONS 9, 164
COMPENSATION CAPACITANCE 79
COMPENSATION RESISTOR 69, 70
COMPONENTS 31, 80, 183, 250
COMPRESSION AMPLIFIER 169
COMPRESSION AMPLIFIER CIRCUIT 169
COMPUTER 51, 105, 252
COMPUTER CIRCUITS 197
CONDUCTANCE 18
CONDUCTORS 248
CONNECTING PAD 23
CONNECTING WIRES 21
CONNECTIONS 116, 223
CONSTANT CURRENT 86
CONSTANT CURRENT SOURCE (CCS) 86, 136
CONSTANT RATE 224
CONSTRAINT 138
CONSTRUCTION 182
CONSTRUCTION ERROR 31
CONTINUOUS WAVE (CW) 187
CONTROL 9
CONTROL VOLTAGE 181
COS/MOS TRANSISTOR ARRAY 175
COSINE HARMONICS 207
COSINE WAVE 228
COST 4
COUNTER-VOLTAGE 74
COUPLING PATH 122
CRITICAL FREQUENCIES 187
CURRENT 5, 8, 28, 37, 53, 69, 70, 143, 149, 178, 248, 251
CURRENT AMPLIFIER 7
CURRENT DIFFERENCE AMPLIFIER (CDA) 4, 17, 135, 137, 150
CURRENT GAIN 129, 130
CURRENT LEVELS 115
CURRENT LOAD 122
CURRENT LOADING IA 117
CURRENT LOADING ISOLATION AMPLIFIER 122
CURRENT MIRROR 144

CURRENT MIRROR DYNAMIC
RESISTANCE 138
CURRENT MIRROR RESISTANCE 138
CURRENT MIRROR TRANSISTOR
136, 138
CURRENT RATINGS 245
CURRENT SOURCE TRANSMITTER 95
CURRENT VARIATION 122
CURRENT-LIMITING RESISTORS 21
CURRENT-OUTPUT DEVICE 133
CURRENTS 68
CUT-OFF FREQUENCY 190
CYCLE 119

D

DAC 216
DAC0806 214
DARLINGTON 99
DARLINGTON TRANSISTORS 14
DATA SHEETS 137
DC 31, 46, 59, 107, 243
DC AMPLIFIER CIRCUITS 15
DC AMPLIFIERS 4, 103, 118
DC BALANCE 109, 110
DC BALANCE CONTROL 109
DC BATTERY 135
DC BLOCKING CAPACITOR 171
DC COUPLING 105
DC DIFFERENTIAL AMPLIFIER
17, 87, 88, 97, 98, 100, 102, 125, 237
DC DIFFERENTIAL AMPLIFIER CIRCUIT
90, 93, 95
DC DIFFERENTIAL CIRCUIT 90, 166
DC HALF-CELL POTENTIAL 107
DC LEVEL 12, 239
DC OFFSET 68, 214, 228
DC OFFSET BIAS 159
DC OFFSET COMPONENT 231
DC OFFSET POTENTIAL 12, 229
DC OFFSET VOLTAGE 60, 229
DC OFFSETS 67
DC OUTPUT 26
DC OUTPUT OFFSET 26, 68
DC POWER 123, 177, 252
DC POWER CONNECTION 152
DC POWER DISTRIBUTION 246
DC POWER LINES 78, 152, 250
DC POWER SUPPLY 11, 12, 15, 21,
58, 77, 89, 112, 113, 115, 116,
118, 122, 135, 141, 164, 165,
170, 243, 247, 253
DC POWER SUPPLY CIRCUITS 58, 243
DC POWER SUPPLY CONNECTIONS
47, 159
DC POWER SUPPLY LINE 20
DC POWER SUPPLY RAIL 89
DC POWER SUPPLY RIPPLE REDUCTION
78
DC POWER SUPPLY TERMINALS 20
DC POWER SUPPLY VOLTAGES 89
DC REPRESENTATION 238
DC RETURN LINES 250
DC SUPPLY POWERED 116
DC SUPPLY VOLTAGES 30
DC VOLTAGE 68, 224
DC VOLTAGE LEVEL 113, 119
DC VOLTMETER 91, 108
DC-COUPLED 55
DC-COUPLED OSCILLOSCOPE 108
DC-TO-DC CONVERTER 11, 121, 122, 126
DECADE 186
DECIBELS 84, 186
DECIBELS-PER-DECADE 186
DECOUPLING BYPASS CAPACITORS 89
DECOUPLING CAPACITORS 165, 177
DEFECTS 26
DEGRADATION 183
DELTA 224
DEMODULATOR 118
DENOMINATOR 100
DERIVATIVE OF BLOOD PRESSURE 226
DESIGN 69, 88, 141, 152
DESIGN EQUATIONS 26
DESIGNER 60
DEVICES 14, 25, 29
DIASTOLIC 238
DIASTOLIC PRESSURE 226
DIASTOLIC VALUE 239
DIE 21
DIE SIZES 21

DIFET 100
DIFFERENTIAL AMPLIFIER 10, 86, 90, 92, 97, 136, 146
DIFFERENTIAL AMPLIFIER CONFIGURATION 94
DIFFERENTIAL INPUT CIRCUIT 129
DIFFERENTIAL INPUT IMPEDANCE 147
DIFFERENTIAL INPUT SIGNAL 100
DIFFERENTIAL INPUT STAGE 86
DIFFERENTIAL INPUTS 26, 32, 36, 83, 84, 94
DIFFERENTIAL OUTPUT 148
DIFFERENTIAL SIGNALS 85
DIFFERENTIAL VOLTAGE GAIN 85, 126, 146
DIFFERENTIATED WAVEFORM 232
DIFFERENTIATOR 221, 223, 224, 232
DIFFERENTIATOR CIRCUIT 231
DIGITAL 183
DIGITAL CAPACITANCE METERS 191
DIGITAL CLOCK CIRCUIT 197
DIGITAL CLOCKS 197
DIGITAL DATA LOGGER 105
DIGITAL DC VOLTAGE READING INSTRUMENT 239
DIGITAL DEVICES 4
DIGITAL ELECTRONICS TECHNOLOGY 214
DIGITAL INTEGRATED CIRCUIT 203
DIGITAL LOGIC 197
DIGITAL MULTIMETERS 191
DIGITAL QUANTIZATION 216
DIGITAL SIGNALS 252
DIGITAL VOLTMETER 108
DIGITAL WAVEFORM GENERATORS 214
DIGITAL-TO-ANALOG CONVERTER (DAC) 214
DIODE 20, 21, 211, 234, 235, 246
DIODE CONNECTED 136
DIODE CURVE 240
DIP IC PACKAGE 121
DIP PACK 23
DIP PACK CONSTRUCTION 23
DIP PACKAGE 110, 125, 195
DIRECT CURRENT (DC) 243, 246
DIRECT METHOD 121
DISC CERAMIC CAPACITORS 77
DISCREET TRANSISTOR 4
DISCRETE OPERATIONAL AMPLIFIERS 110
DISPLAY WIRING 51
DISTRIBUTED CIRCUIT CAPACITANCES 203
DISTRIBUTED VOLTAGE REGULATION 253
DOCTORS 226, 239
DP/DT 238
DRIFT 4, 15
DRIVE POWER 50
DRIVING CIRCUIT 43
DUAL IN-LINE PACKAGES (DIP) 23
DUAL POTENTIOMETER 189
DUAL POWER SUPPLY 11
DUAL-OPERATIONAL AMPLIFIER 153
DUAL-POLARITY POWER SUPPLY CIRCUITS 157
DUAL-SECTION VARIABLE CAPACITOR 189
DYNAMIC RESISTANCE 136, 144

E

EARPHONES 170
ECG AMPLIFIER 106, 107, 108
ECG PEAK SIGNAL 107
ECG PREAMPLIFIER 108
ECG SIGNAL 106, 107, 108
ECG WAVEFORM 107
EFFICIENCY 6
ELECTRIC CURRENT 23
ELECTRICAL RESISTANCE 106
ELECTRICAL SIGNAL 106
ELECTROCARDIOGRAPH 106
ELECTRODES 106, 107, 198
ELECTROLYTIC CAPACITOR 20, 152
ELECTROLYTIC SKIN 107
ELECTROLYTICS 20, 90
ELECTRONIC BOOKS 140
ELECTRONIC CIRCUIT 29, 187, 243, 247
ELECTRONIC DEVICE 25

ELECTRONIC DIFFERENTIATORS 231
ELECTRONIC INSTRUMENTATION CIRCUITS 221, 224
ELECTRONIC INTEGRATOR 227, 228
ELECTRONIC INTEGRATOR CIRCUITS 227
ELECTRONIC NOISE 30
ELECTRONIC OPTOCOUPLERS 120
ELECTRONIC PARTS DISTRIBUTORS 92
ELECTRONIC SIGNALS 228
ELECTRONIC SWITCH 118, 237
ELECTRONIC WAVEFORMS 197
ELECTRONICS 45, 197
ELECTRONS 30
ELECTROPHORESIS SYSTEM 115
EMITTER TERMINALS 86
EMITTERS 118
ENVELOPE 118
EPROM 217
EQUATION 67, 75, 140, 143, 144, 147
ERASABLE PROGRAMMABLE READ-ONLY MEMORY (EPROM) 217
ERROR 228
ESAKI DIODE 201
EXAMPLE 60
EXAR 214
EXAR XR-2206 213
EXTERNAL CIRCUIT 28
EXTERNAL RESISTOR NETWORK 135

F

FEEDBACK 26, 77, 198
FEEDBACK AMPLIFIERS 49
FEEDBACK FACTOR 50, 79
FEEDBACK NETWORK 26, 50, 53, 69, 155, 157, 201, 202, 207
FEEDBACK OSCILLATOR CIRCUITS 201
FEEDBACK OSCILLATORS 198, 201
FEEDBACK RESISTANCE 94, 162
FEEDBACK RESISTOR 38, 39, 42, 44, 45, 46, 69, 80, 148
FEEDBACK SIGNAL 203
FET 118
FIELD EFFECT TRANSISTORS 28
FIGURE OF MERIT 187
FILTER 95, 119, 185, 191, 192, 193
FILTER FREQUENCY RESPONSE 185
FINITE INPUT CURRENT 28
FIXED GAIN 15
FIXED RESISTOR 100, 103
FM DETECTOR 120
FM RECEIVER 30
FOIL 179
FOLLOWER 10
FORWARD BIASED 21
FORWARD GAIN 207
FREQUENCY 90, 103, 139, 159, 166, 168, 186, 187, 189, 203, 208
FREQUENCY COMPENSATION 75
FREQUENCY LIMIT 55
FREQUENCY OF OPERATION 146
FREQUENCY RANGE 76
FREQUENCY RESPONSE 39, 44, 67, 76, 104, 106, 144, 154, 157, 168, 185
FREQUENCY RESPONSE CURVE 48, 149
FREQUENCY RESPONSE LIMIT 45
FREQUENCY RESPONSE LIMITATIONS 76
FREQUENCY RESPONSE TAILORING 39
FREQUENCY SELECTIVE FILTER 185
FUNCTION GENERATOR CHIP 213
FUNDAMENTAL FREQUENCY 207

G

GAIN 7, 31, 46, 53, 60, 76, 79, 97, 98, 100, 125, 143, 146, 155, 156, 162, 171, 182, 183, 185, 187, 201, 207, 211
GAIN BANDWIDTH 77
GAIN BANDWIDTH PRODUCT (GBP) 76, 77
GAIN CONTROL 100
GAIN EQUATION 100
GAIN EXPRESSION 101
GAIN FREQUENCY 77
GAIN OF 100 47
GAIN OF 100 PREAMPLIFIER 46
GAIN OF TEN PREAMPLIFIER 45
GAIN RANGE 117
GAIN RESISTORS 142

GAIN-BANDWIDTH PRODUCT (GBP) 15, 31, 67, 78, 156, 180
GAIN-OF-100 AC AMPLIFIER 141
GAIN-OF-100 NONINVERTING AMPLIFIER 55
GAIN-OF-24 ISOLATED INPUT PREAMPLIFIER 122
GBP 31, 76, 180
GENERAL PURPOSE OP AMPS 14
GENERAL WIDEBAND PREAMPLIFIER 175
GOLD 23
GRADES 14
GRAPHIC EQUALIZERS 159
GRAPHS 80
GREATER-THAN-UNITY GAIN 49, 51
GREEK LETTER 224
GROUND 12, 32, 35, 59, 74, 75, 77, 116, 118, 150, 183, 250
GROUND LOOPS 78, 250
GROUND POTENTIAL 12, 35
GROUND SYSTEM 125
GROUND TRACKS 183
GROUND/COMMON 125
GUITAR SYSTEMS 162

H

H.H. SCOTT 3
H11 169
HALF-CELL POTENTIAL 107
HARMONICS 203
HEART 106
HEAT SINK 31, 173, 245
HEAT TRANSFER GREASE 173
HERTZ 60
HEURISTICS 183
HF SHORTWAVE 177
HIGH FREQUENCIES 164
HIGH GAIN CIRCUITS 68
HIGH IMPEDANCE 115
HIGH IMPEDANCE SOURCE 52
HIGH INPUT IMPEDANCE 47, 165
HIGH POWER AMPLIFIERS 151
HIGH VALUE CAPACITORS 89
HIGH VOLTAGE 16
HIGH-PASS CASE 192
HIGH-PASS FILTER (HPF) 186, 190, 231, 233
HIGHER VALUE CAPACITORS 20
HISS SIGNALS 242
HOLDING POTENTIAL 198
HOSPITAL 115
HOT LINE 250
HUMAN HEARING 151
HYBRID 16

I

I/O 253
IA 100
IA CIRCUIT 100, 108
IBM PC 252
IC 4, 170, 175, 177, 198
IC AMPLIFIER 20, 152, 202
IC AUDIO POWER AMPLIFIERS 170
IC AUDIO PREAMPLIFIER 152
IC CHIPS 3, 31
IC CONSTRUCTION 4
IC DEVICE 23
IC FILTERS 195
IC LINEAR AMPLIFIERS 127
IC OPERATIONAL AMPLIFIERS 4
IC OPTOISOLATORS 121
IC PACKAGE 21, 99
IC PREAMPLIFIERS 152
IC VOLTAGE COMPARATORS 14
ICIA 95
IDEAL AMPLIFIER 75
IDEAL DEVICES 67
IDEAL OP AMP 53, 166
IDEAL OPERATIONAL AMPLIFIER 36
IEEE STANDARD 9
IF AMPLIFIER FILTER 187
IMPEDANCE 15, 26, 47, 90, 97, 177
IMPEDANCE TRANSFORMATION 50, 51, 54
INA-102 110
INA-117 110, 111
INDUCTANCES 77
INDUCTORS 185

INDUSTRIAL INDUCTIVE HEATING 6
INDUSTRY STANDARD 92
INDUSTRY STANDARD OPERATIONAL AMPLIFIER 130
INFINITE BANDWIDTH 31
INFINITE INPUT IMPEDANCE 27
INFINITE OPEN-LOOP GAIN 26
INPUT 26, 36, 92, 100, 178, 181, 233
INPUT RESISTOR 38
INPUT AMPLIFIERS 99
INPUT BIAS 154
INPUT BIAS CURRENT 37, 87, 97
INPUT CAPACITANCE 81
INPUT CAPACITANCE COMPENSATION 80
INPUT CAPACITOR 104, 142, 144
INPUT CIRCUIT 28, 77, 104, 126, 129, 136, 168
INPUT COUPLING CAPACITORS 112
INPUT CURRENT 14
INPUT CURRENT 7, 14, 27
INPUT FUNCTION 27, 36
INPUT IMPEDANCE 26, 27, 43, 51, 54, 57, 60, 87, 90, 97, 98, 106, 122, 142, 143, 144
INPUT NOISE VOLTAGE 30
INPUT OFFSET CURRENT 68, 69, 70
INPUT OFFSET VOLTAGE 69, 70
INPUT PINS 9
INPUT POTENTIAL 7
INPUT REFERENCE CURRENT 138
INPUT RESISTANCE 29, 44, 81, 140, 147
INPUT RESISTOR 38, 43, 59, 69, 97, 104, 138, 146, 147, 148, 162
INPUT SIGNAL 9, 30, 35, 49, 68, 83, 118, 122, 129, 203, 228
INPUT SIGNAL CABLES 84
INPUT SIGNAL PATH 60
INPUT SIGNAL POTENTIAL 37
INPUT SIGNAL SOURCE 125
INPUT SIGNAL VOLTAGE 7, 165
INPUT SIGNALS 15
INPUT STAGE 14, 86
INPUT TRANSISTORS 69
INPUT TRIGGER PULSE 218
INPUT VOLTAGE 7, 27, 129, 235, 240
INPUT WAVEFORM 211, 227
INPUTS 29, 32
INSTRUMENT 238
INSTRUMENTATION 9, 221, 237
INSTRUMENTATION AMPLIFIER (IA) 17, 76, 97, 100, 102, 104
INSTRUMENTATION AMPLIFIER APPLICATIONS 105
INSTRUMENTATION AMPLIFIER CIRCUIT 98
INSTRUMENTATION APPLICATIONS 95
INSTRUMENTATION CIRCUIT 51
INTEGRAL 226
INTEGRATED CIRCUIT (IC) 3, 21, 25, 121, 151, 152
INTEGRATED CIRCUIT INSTRUMENTATION AMPLIFIERS (ICIA) 95
INTEGRATED CIRCUIT OPERATIONAL AMPLIFIER 202
INTEGRATOR 212, 223, 229, 231
INTEGRATOR ACTION 228
INTEGRATOR SERVICE 229
INTEGRATORS 198, 221, 224
INTENSIVE CARE 226
INTERCONNECTION 165
INTERNAL CIRCUITRY 15
INTERNAL IMPEDANCE 51
INTERNAL NOISE 30
INTERNAL RESISTANCE 28
INTERNAL RESISTORS 4
INTERNAL VOLTAGE SOURCE 69
INVERTING AMPLIFIER 47, 79, 143
INVERTING AMPLIFIER CIRCUITS 35
INVERTING CASE 144
INVERTING FOLLOWER 9, 32, 35, 36, 43
INVERTING FOLLOWER AMPLIFIERS 43
INVERTING FOLLOWER CIRCUIT 35, 43, 53, 138
INVERTING INPUT 9, 32, 35, 75, 83, 146, 172
INVERTING INTEGRATOR 227
INVERTING SCHMITT TRIGGER 209
IONIZATION 198

ISO122P 125
ISOLATED AMPLIFIER 122
ISOLATED "B" SIDE 116
ISOLATED GROUND 116, 125
ISOLATED POWER SUPPLY 116
ISOLATED SECTION 117
ISOLATED STAGES 121
ISOLATION 51, 116, 118, 122
ISOLATION AMPLIFIER 113, 115, 116, 117, 122, 123, 127
ISOLATION MODE REJECTION 125

J

JAMECO 213, 214
JFET 14, 169, 202
JFET INPUT DEVICES 28
JFET INPUT OPERATIONAL AMPLIFIER 122
JFET INPUT TRANSISTORS 90
JFET TRANSISTORS 69, 87
JUNCTION FIELD EFFECT TRANSISTOR (JFET) 14, 169, 202
JUNCTION POTENTIAL 233

K

KCL 38, 53, 86
KIRCHOFF'S CURRENT LAW (KCL) 38, 53, 86
KNEE 240
KNEE FREQUENCY 227, 231
KNOB 163

L

LAB EXPERIMENTS 25
LABORATORY AMPLIFIER 120
LAMP 199
LARGE SCALE INTEGRATION (LSI) 4
LAYOUT 77, 78, 183
LC NETWORK 203
LEADS 23, 44, 183
LEAKAGE 28
LEAKAGE CURRENTS 87, 115
LED 121, 122, 169
LED BIAS 122
LIGHT-EMITTING DIODE (LED) 120
LIGHTNING IMPULSES 242
LIMITED/SPECIAL PURPOSE DEVICES 16
LINE DRIVER AMPLIFIER 148, 166
LINE DRIVERS 165
LINE RECEIVER 165, 147
LINEAR AMPLIFIERS 20
LINEAR ANALOG DEVICES 4
LINEAR DATA BOOK 152
LINEAR IC AMPLIFIER 18, 135
LINEAR IC CIRCUITS 243
LINEAR INTEGRATED CIRCUITS 229
LINEAR OPERATION 6
LINEAR VARIABLE DIFFERENTIAL TRANSFORMER (LVDT) 238
LINEAR VOLTAGE RAMP 212
LM-1303 152, 159
LM-2002 170, 172
LM-2002 IC AUDIO POWER AMPLIFIER 171
LM-301 152, 165
LM-320N-XX 244
LM-340N-XX 244
LM-340T-5 244
LM-381 152, 153, 154
LM-382 152
LM-386 170, 172
LM-386 AUDIO PREAMPLIFIER 170
LM-3900 136, 150
LOAD 5, 28, 51
LOAD IMPEDANCE 143, 183
LOAD RESISTANCE 29, 130, 143
LOAD RESISTOR 122
LONG CABLE 51
LOOP GAIN 26, 203
LOUDSPEAKER 7, 167, 170
LOUDSPEAKERS 151
LOW DRIFT 15
LOW FREQUENCY 180
LOW IMPEDANCE LOAD 52
LOW IMPEDANCE POWER SOURCE 51
LOW INPUT CURRENT 14
LOW NOISE 14
LOW POWER 15
LOW-GAIN OP AMP CIRCUIT 68

LOW-LEAKAGE 235
LOW-PASS CASE 191
LOW-PASS FILTER 187, 190, 211
LOW-PASS OUTPUT 192
LOW-PRESSURE INERT GAS 198
LVDT 238

M

MANUFACTURERS 32, 79, 87, 117
MAP 226, 239
MAR-1 177, 179
MAR-1 VOLTAGE 178
MAR-1 WIDEBAND PREAMPLIFIER 177
MAR-X 177
MARKET 14
MASTER GAIN CONTROL 162
MC-1303 159
MEAN ARTERIAL PRESSURE (MAP) 226, 238
MEAN VALUE DETECTOR 240
MEASUREMENT 9
MEDICAL BLOOD PRESSURE MONITORS 226
MEDICAL DIATHERMY 6
MEDICAL ELECTRONICS 226
MEDICAL EQUIPMENT 115
MEDICAL INSTRUMENTS 187
MEDICAL MONITORING DEVICE 122
MEDIUM SCALE INTEGRATION (MSI) 4
METAL OXIDE SEMICONDUCTOR FIELD EFFECT TRANSISTOR 202
METALLIC ELECTRODES 107
METER INPUT 45
MHO 8
MICA 236
MICROAMPERES 7
MICROCOMPUTER 251
MICROFARADS 60
MICROPHONE 151, 162, 167
MICROPHONE PREAMPLIFIER 76, 167
MICROSIEMENS 8
MICROWAVE TRANSISTOR 177
MIDBAND 59
MIDBAND FREQUENCY 187
MILLER INTEGRATOR 195, 212, 240
MILLER INTEGRATOR CONFIGURATION 112
MILLIAMPERES 7
MILLISIEMENS 8
MILLIVOLT 106
MINICIRCUITS CATALOG 178
MINIDIP PACKAGE 93, 107, 180
MINIMUM VALUE DETECTOR 241
MIXER 165
MODERN DESIGN 121
MODULATOR 118, 120
MODULATOR STAGE 118
MONOPOLAR DC POWER SUPPLY 15
MONOSTABLE CIRCUIT 218
MONOSTABLE MULTIVIBRATOR 217, 232
MORSE CODE 187
MOSFET 14, 28, 202
MOSFET INPUT TRANSISTORS 90, 229
MOSFET TRANSISTORS 69, 87
MOTHERBOARD 252, 253
MOTOROLA 159
MOUNTING TAB 173
MULTIPLE AMPLIFIER 21
MULTIPLE CDA INTEGRATED CIRCUIT 150
MULTIPLE DEVICES 16
MULTIPLE FEEDBACK PATH CIRCUIT 192
MULTIPLE FUNCTION FILTER 192
MULTIPLE OP AMP CIRCUIT 44
MULTIPLE OUTPUTS 51
MULTIPLE-INPUT INVERTING FOLLOWERS 42
MULTISTAGE CIRCUITS 77
MULTIVIBRATOR 197
MUSIC 162

N

NAB-COMPENSATED TAPE PREAMPLIFIERS 157
NARROW-BAND DEVICES 175
NATIONAL SEMICONDUCTOR 136, 152
NATURAL NOISE 30
NEGATIVE COMMON RETURN 135
NEGATIVE FEEDBACK 26, 49
NEGATIVE FEEDBACK AMPLIFIER 26

NEGATIVE FEEDBACK CIRCUIT 159
NEGATIVE FEEDBACK PATH 211
NEGATIVE FEEDBACK RESISTOR 146
NEGATIVE INPUT 234
NEGATIVE REGULATORS 12
NEGATIVE RESISTANCE DEVICES 198, 200, 201
NEGATIVE RESISTANCE PROPERTY 200
NEGATIVE RESISTANCE ZONE (NRZ) 201
NEGATIVE-GOING OUTPUT SIGNAL 83
NEGATIVE-GOING OUTPUT VOLTAGE 227
NEON GLOW-LAMPS 198
NEON LAMP 199
NO BLEED RESISTOR 236
NOISE 14, 19, 20, 26, 30, 138, 144
NOISE CONTRIBUTION 26
NOISE SIGNALS 30
NOISE VOLTAGES 30
NOISY REFERENCE SOURCE 144
NON-CHIP CAPACITORS 179
NON-COMPENSATED OP AMPS 31
NON-IDEAL OPERATIONAL AMPLIFIERS 26
NON-IDEAL OUTPUT 29
NON-ISOLATED "A" SIDE 116
NON-LINEARITY 125
NON-OPERATIONAL AMPLIFIER 135
NON-SINUSOIDAL WAVEFORM GENERATORS 207
NON-SINUSOIDAL WAVEFORMS 207
NON-ZENER MODE 21
NONINVERTING AMPLIFIER 55
NONINVERTING AMPLIFIER CDA CONFIGURATION 143
NONINVERTING AMPLIFIER CIRCUITS 143
NONINVERTING CDA AMPLIFIER 143
NONINVERTING FOLLOWER 10, 32, 49, 53, 54, 64, 69, 145
NONINVERTING FOLLOWER CIRCUIT 144
NONINVERTING FOLLOWER WITH GAIN 53, 99
NONINVERTING INPUT 9, 12, 32, 35, 37, 49, 53, 62, 74, 83, 86, 104, 136, 143, 150, 168
NONINVERTING INPUTS 146
NONINVERTING OPERATIONAL AMPLIFIER 133
NONINVERTING UNITY GAIN 79
NONLINEAR 138
NONLINEAR KNEE REGION 233
NONSTANDARD VALUE 143
NORTON AMPLIFIER 4, 17, 135, 136
NORTON'S CURRENT THEOREM 135
NOTCH FILTER 187, 189
NOTCH FILTER CIRCUIT 187
NOTCH FREQUENCY 189
NPN 28, 202
NPN BIPOLAR DEVICES 14
NPN DIFFERENTIAL PAIR 68
NPN OUTPUT TRANSISTOR 136
NPN TRANSISTORS 118
NULL CONTROL 122
NURSES 239

O

OA 9
OCTAVE 186
OFFSET CURRENTS 68, 69
OFFSET NULL POTENTIOMETER 71, 230
OFFSET VOLTAGE 71, 73, 75
OHM RESISTOR 233
OHMMETER 88
OHMS 60
OHM'S LAW 8, 37, 53, 69, 133, 143, 198, 200, 201
OP AMP 9, 14, 15, 26, 27, 28, 29, 31, 37, 51, 54, 65, 77, 83, 87, 89, 92, 104, 135, 157, 159, 161, 211, 212, 221, 223, 235
OP AMP CIRCUITS 165, 243
OP AMP CONFIGURATION 35, 49
OP AMP DIFFERENTIATOR 239
OP AMP FEEDBACK RESISTOR 133
OP AMP INPUT 53, 68
OP AMP INVERTING FOLLOWER 38

OP AMP INVERTING FOLLOWER CIRCUIT 37
OP AMP OUTPUT 234
OP AMP OUTPUT VOLTAGE 50
OP AMP SYMBOL 17
OP AMPS 14, 26, 30, 31, 99, 195, 221
OPA-2662 180
OPA-620 180
OPA-621 180
OPA-622AP 180
OPA-622AU 180
OPA-623 180
OPA-627 180
OPA-637 180
OPA-678 180
OPEN-LOOP GAIN 26, 50
OPEN-LOOP VOLTAGE GAIN 26
OPERATING FREQUENCY 205
OPERATING WAVEFORMS 198
OPERATIONAL AMPLIFIER (OA, OP AMP) 3, 8, 9, 11, 14, 18, 28, 29, 30, 32, 35, 36, 43, 44, 67, 69, 71, 77, 80, 83, 84, 90, 98, 99, 108, 129, 135, 138, 145, 152, 154, 157, 162, 164, 165, 166, 180, 189, 203, 209, 227, 229, 233, 235
OPERATIONAL AMPLIFIER IC DEVICES 26
OPERATIONAL AMPLIFIER PREAMP 154
OPERATIONAL TRANSCONDUCTANCE AMPLIFIER (OTA) 4, 18, 129, 130
OPTICAL COUPLING 121
OPTICALLY-COUPLED CIRCUITS 120
OPTICALLY-COUPLED IA 117
OPTICALLY-COUPLED ISOLATION AMPLIFIERS 121
OPTOISOLATOR 120, 122, 169
OSCILLATION 19, 20, 26, 77, 81, 183, 198, 203, 206, 207, 222, 223
OSCILLATION FREQUENCY 78
OSCILLATOR 51, 197
OSCILLATOR CIRCUIT 203
OSCILLOGRAPHIC PAPER RECORDER 239
OSCILLOSCOPE 51, 72, 92, 105, 106, 108, 223, 239
OTA 129, 133, 180
OUTPUT 5, 26, 29, 67, 83, 118, 119, 146, 171, 178, 233
OUTPUT AMPLITUDE 205
OUTPUT BUFFER 125
OUTPUT CAPACITOR 143
OUTPUT CONNECTION 51
OUTPUT CURRENT 5, 129, 133
OUTPUT FUNCTION 27, 36
OUTPUT IMPEDANCE 26, 28, 29, 51, 134, 166
OUTPUT INDICATOR 92
OUTPUT LEVEL CONTROL 211
OUTPUT LOAD RESISTOR 60
OUTPUT NOISE VOLTAGE 30
OUTPUT OFFSET ERROR 28
OUTPUT OFFSET POTENTIAL 31
OUTPUT POTENTIAL 37
OUTPUT POWER 7
OUTPUT PULSE 218
OUTPUT RESISTANCE 29, 131, 132
OUTPUT SIGNAL 9, 30, 35, 49, 83, 85, 211
OUTPUT SIGNAL FLOWING 77
OUTPUT SIGNAL POTENTIAL 37
OUTPUT TERMINAL 49, 68, 139, 143, 145
OUTPUT VOLTAGE 29, 31, 73, 83, 110, 133, 139, 145, 234
OUTPUT VOLTAGE OFFSET 70
OUTPUT VOLTAGE SHIFT 110
OUTPUT VOLTAGE TRANSFER FUNCTION 42
OUTPUTS 211

P

PACKAGE PINS 21
PACKAGE STYLE 23, 44
PACKAGES 23
PANEL 163
PAPER STRIP-CHART RECORDER 106
PARALLEL COMBINATION 71

PASSBAND FREQUENCY RESPONSE 185
PASSIVE FILTERS 185
PATIENT 115
PCB 179, 253
PEAK DETECTOR 235, 236
PEAK DETECTOR CIRCUITS 235, 236
PEAK HOLDER 236
PEAK VALUE 106
PERIODIC 197
PERIODIC WAVEFORM 197
PHASE 183
PHASE DETECTOR 120
PHASE SHIFT NETWORK 206
PHASE SHIFT OSCILLATORS 206
PHASE-LOCKED LOOP (PLL) 120
PHONOGRAPH CARTRIDGES 151
PHONOGRAPH PREAMPLIFIERS 157
PHOTOLITHOGRAPHIC 21
PHOTORESISTOR 121
PHOTOTRANSISTOR 121
PIEZORESISTIVE 238
PIEZORESISTIVE BRIDGES 111
PIN 23, 179
PIN-OUT DESIGNATIONS 125
PINOUTS 89
PINS 95
PLASTIC 23
PLUG-IN CARDS 253
PLUG-IN CONNECTORS 253
PN JUNCTIONS 46
PNP 202
PNP BIPOLAR DEVICES 14
PNP DRIVER TRANSISTOR 136
PNP TRANSISTORS 28
POLARITIES 83, 119
POLARITY GAIN 84
POLYCARBONATE 236
POLYSTYRENE 236
PORTABLE EQUIPMENT 19
POSITION 109
POSITIVE FEEDBACK 26
POSITIVE RESISTANCE 200
POSITIVE-GOING INPUT SIGNAL 83
POSITIVE-GOING INPUT VOLTAGE 227
POSITIVE-GOING OUTPUT SIGNAL 83
POST AMPLIFIER CIRCUIT 109
POTENTIAL 11, 176
POTENTIOMETER 71, 72, 73, 91, 93, 100, 102, 103, 107, 163, 230
POTENTIOMETER WIPER 71, 74
POTS 189
POWER 15, 51, 209
POWER AMPLIFIER 170
POWER CONSUMPTION 15, 130
POWER DISTRIBUTION SCHEMES 252
POWER DISTRIBUTION SYSTEM 247
POWER DRIVER 51
POWER GAIN 51
POWER SOURCE 51
POWER SUPPLY 19, 116, 118, 122, 177, 181, 246, 250
POWER SUPPLY CARRIER SIGNAL 120
POWER SUPPLY COMMON 32
POWER SUPPLY LINES 21, 116, 183
POWER SUPPLY PINS 46
POWER SUPPLY REJECTION 19
POWER SUPPLY RIPPLE REDUCTION 253
POWER SUPPLY SENSE METHOD 253
POWER SUPPLY SENSITIVITY (PSS) 250
POWER SUPPLY TERMINALS 19, 21, 115
POWER TERMINALS 251
POWER TRANSISTOR 171
PRACTICAL CIRCUITS 69, 87, 144
PRACTICAL INSTRUMENTATION APPLICATION 237
PRACTICAL INTEGRATOR 229
PREAMPLIFIER CIRCUITS 167
PREAMPLIFIERS 5, 151, 152, 157, 168
PRECISE RECTIFIERS 233, 235
PREREGULATION 253
PRESSURE 226
PRESSURE DERIVATIVE (DP/DT) 226
PRESSURE WAVEFORM 239
PRESSURE WAVEFORM DERIVATIVE 239
PRINTED CIRCUIT BOARD 92, 179, 183, 252
PRINTED CIRCUIT PATTERN 92, 179
PRINTED CIRCUIT TRACES 183
PRINTED CIRCUIT TRACKS 252

PROBE 232
PROCESSOR CARD 253
PROJECT 45, 247
PROPERTIES 26, 36
PROTECTING LINEAR IC AMPLIFIERS 18
PROXIMITY DETECTORS 221
PSS 250
PSS SPECIFICATION 250
PUBLIC ADDRESS SYSTEMS 162
PULSE 182, 197
PULSE-COUNTING FM DETECTOR 120
PUSH-PULL 119

Q

Q 187
QUADRATURE 228
QUADRATURE DETECTOR 237
QUADRATURE PHASE DETECTOR 237
QUADRATURE SIGNAL 231
QUADRATURE SINE WAVE SIGNALS 211
QUIESCENT CURRENT 122

R

RADIO FREQUENCY (RF) 254
RADIO FREQUENCY CHOKES (RFCS) 254
RADIO STATION 162
RADIO TRANSMITTERS 6
RAMP GENERATOR CIRCUIT 212
RATE OF CHANGE 224
RC CIRCUIT 205
RC HIGH-PASS FILTER 231
RC INTEGRATOR 232
RC NETWORK 165, 203, 228, 232
RC TIME-CONSTANT 198, 203, 231, 232, 236
RCA CA-3140 47
REACTANCE 62, 75
RECORD PLAYER 162
RECORDER 106
RECTIFIER 233, 240
REFERENCE CURRENT 143, 144
REFERENCE RESISTOR 150
REFERENCE SIGNAL 119
REFERENCE SOURCE 144
REFERENCE VOLTAGE 143
REFERENCE VOLTAGE SOURCE 138
REGULATORS 245
REJECTION RATIO ADJUSTMENT 237
RELAXATION OSCILLATORS 198, 201
REMOTE PREAMPLIFIER 165
REPRESENTATIVE APPLICATIONS 152
RESEARCH INSTRUMENT 238
RESET TIME 213
RESISTANCE 51, 139
RESISTANCE ELEMENT 169, 181
RESISTANCE NETWORK 203
RESISTANCE VALUES 103
RESISTIVE CIRCUITS 30
RESISTIVE NOISE 30
RESISTOR 30, 38, 41, 53, 57, 59, 60, 69, 86, 91, 100, 102, 103, 119, 133, 138, 149, 155, 157, 166, 192, 229, 233
RESISTOR CAPACITOR (RC) 227
RESISTOR NETWORKS 30, 135
RESISTOR VOLTAGE DIVIDER 12
RESISTOR VOLTAGE DIVIDER CIRCUIT 144
RESISTOR-CAPACITOR (RC) 231
RESISTOR-CAPACITOR NETWORK 217
RESISTORS 17, 35, 61, 73, 87, 88, 90, 94, 97, 104, 109, 111, 167, 185, 189, 195, 206, 214, 223
RESOLUTION 91
RESONANT FREQUENCY 77
RETURN LINES 250
REVERSE POLARITY 21
REVERSE POLARITY CONDITIONS 20
REVERSE POLARITY DC POWER SUPPLIES 19
REVERSE RESISTANCE 235
REVERSE-BIAS 234
RF AMPLIFIERS 183
RF CIRCUITS 175
RF POWER AMPLIFIER CIRCUITS 6
RF SYSTEMS 177
RF/VIDEO 183
RIAA-COMPENSATED PHONOGRAPH PREAMPLIFIERS 157
RISE TIME 242

RLC NETWORKS 159
ROLLED-OFF SEGMENT 157
ROOM TEMPERATURE 137
ROOT MEAN SQUARE (RMS) 240

S

SAWTOOTH 197, 217
SAWTOOTH GENERATOR CIRCUIT 212
SAWTOOTH WAVE 208
SAWTOOTH WAVEFORM 212, 213
SCHEMATIC 122
SCHMITT TRIGGER 209
SCIENTIFIC INSTRUMENTS 187
SCR-LIKE DEVICES 198
SCREWDRIVER 72
SECONDARY WINDING 244
SELF-OSCILLATION 171
SEMICONDUCTOR 4, 9, 200
SEMICONDUCTOR DEVICES 30
SEMICONDUCTOR DIE 23
SENSE LINES 253
SENSE METHOD 253
SERIES 233
SERIES RESISTANCE 166
SERIES RESISTOR 147
SHORT CIRCUITS 51
SHUNT 39
SHUNT RESISTOR 230
SID 241
SIEMEN 8, 18, 129
SIGNAL 26, 31, 49, 51, 55, 75, 83, 90, 103, 118, 120, 122, 151, 164, 167, 181, 211, 214, 224, 236, 239
SIGNAL GENERATOR 108
SIGNAL INPUTS 116
SIGNAL PATHS 118, 250
SIGNAL POTENTIALS 10
SIGNAL PROCESSING CHAIN 118
SIGNAL PROCESSING STAGES 238
SIGNAL SOURCE 51
SIGNAL VOLTAGE 233
SIGNETICS 5534 180
SILICON 21
SILICON DIODES 233
SILICON VALLEY 4
SILICONE 173
SILVER-SILVER CHLORIDE 106
SIMPLE IA CIRCUIT 98
SINE HARMONICS 207
SINE WAVE 197, 211, 214, 228
SINE WAVE GENERATORS 203
SINE WAVE OSCILLATORS 203
SINE WAVES 211, 231, 240
SINGLE DC SUPPLY 15
SINGLE POLARITY 20
SINGLE POLARITY POWER SUPPLY 150
SINGLE SUPPLY OPERATION 11
SINGLE-DEVICE DC DIFFERENTIAL AMPLIFIER 97
SINGLE-ENDED VOLTAGE AMPLIFIER 166
SINGLE-POLARITY DC POWER SUPPLY 153
SINK 53
SKIN SURFACE ELECTRODES 106
SLOPE 186
SNUBBING NETWORK 171
SOLAR STUDIES 241
SOLDER PADS 21
SOURCE 53
SOURCE IMPEDANCE 51, 58, 88, 133, 140
SOURCE RESISTANCE 43, 44
SOURCE VOLTAGE 29
SPAN 109
SPAN CONTROL 109, 110
SPECIFICATIONS 80
SPECTRUM ANALYZER 235
SPLIT-STATOR VARIABLE CAPACITOR 189
SQUARE LAW 240
SQUARE WAVE 197, 208, 209, 211, 212, 214, 228
SQUARE WAVE GENERATOR CIRCUIT 209
SQUARE WAVE GENERATORS 208
SQUARE WAVE SIGNALS 232
STABILITY 221
STANDARD CIRCUIT CONFIGURATIONS 145

STANDARD INSTRUMENTATION
AMPLIFIERS 99
STANDARD RESISTOR VALUES 145
STANDARD VALUE 44, 61, 143, 191
STATE VARIABLE FILTER (SVF) 192
STATIC 242
STAVER V-5 HEAT SINK 173
STEREO 152
STEREO AUDIO 16
STEREO AUDIO PREAMPLIFIER 159
STOPBAND FILTER 187
STRAY CIRCUIT CAPACITANCES 203
STRIP CHART RECORDER 51, 72
STRIP-CHART PAPER RECORDER 105
STRIP-LINE TECHNIQUES 183
STUDENTS 201
SUDDEN IONOSPHERIC DISTURBANCE
(SID) 241
SUMMERS 42
SUMMING JUNCTION 53
SUN 241
SUPER-GAIN AMPLIFIER 145
SUPERBETA 99, 100
SUPERBETA TRANSISTORS 14
SUPPLY VOLTAGE 138, 143
SWITCH 236
SYMBOL 17, 135
SYNCHRONOUS DEMODULATION 118,
119
SYSTOLIC 238
SYSTOLIC OUTPUT VOLTAGE 239
SYSTOLIC PRESSURE 226

T

TANTALUM 90
TANTALUM ELECTROLYTIC CAPACITORS
112
TANTALUM ELECTROLYTICS 20, 152
TAPE HEADS 151, 154
TAPE PLAYER 162
TEE OSCILLATOR 205
TEE-NETWORKS 205
TELEVISION SIGNALS 175
TEMPERATURE 43, 137
TERMINALS 73, 164, 170, 171
TEXTBOOK AMPLIFIERS 67
TEXTBOOK INTEGRATOR CIRCUITS 211
THD 153
THERMAL DRIFT 4, 99
THERMAL ENVIRONMENT 4, 99, 150
THERMISTOR 43
THERMOCOUPLES 111
THERMOSONIC 23
THREE-TERMINAL IC DEVICE 244
THRESHOLD 198
THRESHOLD POTENTIAL 209
TIME 182
TIME AVERAGE 227
TIME CONSTANT 242
TIME-AGC 182
TIME-VARYING VOLTAGE SIGNAL 224
TIMER CIRCUITS 217
TO-220 245
TO-5 METAL TRANSISTOR PACKAGE 23
TO-92 245
TONE CONTROL CIRCUIT 159
TONE CONTROL PREAMPLIFIER STAGE
161
TONE CONTROLS 159
TOTAL HARMONIC DISTORTION (THD)
153
TRACE 223
TRANSCONDUCTANCE 7, 8, 129, 130
TRANSCONDUCTANCE AMPLIFIER
8, 129
TRANSDUCER 43, 221, 238
TRANSDUCER IMPEDANCE 47
TRANSFER EQUATION 36
TRANSFER EQUATION 27, 53
TRANSFER FUNCTION
6, 7, 26, 50, 129, 234
TRANSFORMER 118, 121, 122, 243
TRANSFORMER COUPLED
NONINVERTING AMPLIFIER 63
TRANSIENTS 241, 246
TRANSISTOR 25, 68, 86, 87, 118,
119, 122, 136

TRANSISTOR COLLECTOR 5
TRANSISTOR RADIO BATTERIES 89
TRANSISTOR THEORY 30
TRANSMITTER MODULATION INPUT 162
TRANSMITTER MODULATOR 167
TRANSRESISTANCE 7
TRANSRESISTANCE AMPLIFIER 8
TREBLE 159, 160
TREBLE ROLL-OFF TONE CONTROL 159
TRIANGLE 197
TRIANGLE WAVE 208, 212, 214
TRIANGLE WAVEFORMS 228
TRIANGULAR AMPLIFIER SYMBOL 116
TRIGGER 232
TRIGGER INPUT 232
TRIGGERING THRESHOLD 209
TRIMMER POTENTIOMETERS 214
TTL CHIP 247
TTL DEVICES 251
TUNNEL DIODE 198, 201
TV 9
TV CIRCUITS 175, 183
TWIN-TEE NETWORK 205
TWIN-TEE OSCILLATORS 205

U

UHF 177, 179
UHF SCANNER OPERATION 179
UJTS 198
ULTRASONIC TRANSFORMERS 118
UNIJUNCTION TRANSISTORS (UJT) 198
UNITS 7
UNITS OF CONDUCTANCE 18
UNITY 31, 76, 156, 203
UNITY GAIN 15, 49, 51, 125, 237
UNITY GAIN CONFIGURATION 54
UNITY GAIN FOLLOWER 50, 51
UNITY GAIN INVERTER 236
UNITY GAIN NONINVERTING AMPLIFIER 52
UNITY GAIN NONINVERTING FOLLOWER 50, 236
UNITY GAIN NONINVERTING FOLLOWER CIRCUIT 49
UNIVERSAL DIFF-AMP 92
UNSTABLE 222
USER 115

V

VACUUM TUBE 4, 25, 202
VALUE 44, 62, 144, 191, 224
VALUE CAPACITOR 12
VARIABLE 102
VARIABLE CAPACITORS 195
VARIABLE GAIN DIFFERENTIAL AMPLIFIER 93
VARIABLE GAIN POST-AMPLIFIER STAGE 211
VARIABLE GYRATOR CIRCUIT 189
VARIANT CIRCUIT 71
VCA-610 180, 181
VCO 120
VERY WIDE BANDWIDTH 15
VHF 177, 178
VHF HAM BANDS 179
VIBRATOR SWITCH 237
VIDEO 253
VIDEO AMPLIFIER 175
VIDEO OP AMPS 15
VIRTUAL GROUND 32, 35
VLF 241
VLF BANDS 241
VLF RADIO RECEIVERS 241
VLF RADIO STUDIES 241
VLF SIGNAL 242
VOLTAGE 8, 31, 32, 36, 53, 69, 70, 71, 83, 87, 119, 181, 200, 212, 226, 244, 254
VOLTAGE AMPLIFIER 7, 27, 28, 31, 36, 129, 133, 151
VOLTAGE ANALOG OF PRESSURE 226
VOLTAGE COMPARATORS 14, 198
VOLTAGE CONTROLLED CURRENT SOURCE 95
VOLTAGE CONTROLLED FEATURE 181
VOLTAGE CONTROLLED WIDEBAND AMPLIFIER 180
VOLTAGE DIVIDER 234
VOLTAGE DROP 53, 86, 133, 250

VOLTAGE GAIN 17, 26, 37, 38, 49, 50, 51, 54, 61, 77, 138, 141, 145, 153, 177
VOLTAGE OUTPUT 151
VOLTAGE RATING 246
VOLTAGE REGULATOR 12, 244
VOLTAGE SIGNALS 51, 224
VOLTAGE SOURCE 28, 29, 51
VOLTAGE TRANSFORMATION 166
VOLTAGE VARIATION 122
VOLTAGE-CONTROLLED OSCILLATOR (VCO) 120
VOLTMETER 32, 92, 108
VOLTS 129
VOLUME CONTROL 30, 170

W

WAVE SHAPE 197
WAVEFORM 109, 119, 197, 207, 208, 226, 238, 239
WAVEFORM GENERATOR 51
WAVEFORM GENERATOR CIRCUITS 219
WEIN BRIDGE OSCILLATORS 203
WHEATSTONE BRIDGE 238
WIDE BANDWIDTH 15
WIDEBAND 180, 183
WIDEBAND AMPLIFIER 175, 182, 183
WIDEBAND CHIPS 177
WIDEBAND MODE 180
WIDEBAND OP AMPS 180
WIDEBAND OPERATIONAL AMPLIFIERS 77
WINDING 244
WIPER 73, 181
WIRE 21, 182
WIRE SIZE 250

X

XR-2206 214

Z

ZENER DIODE 20, 74
ZENER POTENTIALS 20, 74
ZERO INPUT IMPEDANCE 28
ZERO INPUT VOLTAGE 31
ZERO NOISE CONTRIBUTION 29
ZERO OUTPUT IMPEDANCE 29
ZERO OUTPUT OFFSET 31

PROMPT® PUBLICATIONS

ES&T Presents TV Troubleshooting & Repair

Electronic Servicing & Technology Magazine

TV set servicing has never been easy. The service manager, service technician, and electronics hobbyist need timely, insightful information in order to locate the correct service literature, make a quick diagnosis, obtain the correct replacement components, complete the repair, and get the TV back to the owner.

ES&T Presents TV Troubleshooting & Repair presents information that will make it possible for technicians and electronics hobbyists to service TVs faster, more efficiently, and more economically, thus making it more likely that customers will choose not to discard their faulty products, but to have them estored to service by a trained, competent profes-ional.

Originally published in *Electronic Servicing & chnology*, the chapters in this book are articles tten by professional technicians, most of whom ice TV sets every day.

Video Technology

226 pages • Paperback • 6 x 9"
BN: 0-7906-1086-8 • Sams: 61086
95 ($25.95 Canada) • August 1996

ES&T Presents Computer Troubleshooting & Repair

Electronic Servicing & Technology

ES&T is the nation's most popular magazine for professionals who service consumer electronics equipment. PROMPT® Publications, a rising star in the technical publishing business, is combining its publishing expertise with the experience and knowledge of *ES&T's* best writers to produce a new line of troubleshooting and repair books for the electronics market. Compiled from articles and prefaced by the editor in chief, Nils Conrad Persson, these books provide valuable, hands-on information for anyone interested in electronics and product repair.

Computer Troubleshooting & Repair is the second book in the series and features information on repairing Macintosh computers, a CD-ROM primer, and a color monitor. Also included are hard drive troubleshooting and repair tips, computer diagnostic software, networking basics, preventative maintenance for computers, upgrading, and much more.

Computer Technology

288 pages • Paperback • 6 x 9"
ISBN: 0-7906-1087-6 • Sams: 61087
$18.95 ($26.50 Canada) • February 1997

CALL 1-800-428-7267 TODAY FOR THE NAME OF UR NEAREST PROMPT PUBLICATIONS DISTRIBUTOR